第一次學做

手工麵包！

U0084832

超安心

序

　　與韋太的認識緣於一個雞肉派餅食譜。話說在2009年，我無意中從網上找尋食譜時，搜尋到韋太網站內的食譜，我就立即嘗試，結果第一次試做就成功了！

　　從此，我便成為韋太的忠實粉絲。

　　我一直都在默默支持韋太，去年（2011年）12月回香港，當然順道要跟韋太學做麵包。

　　（回憶起當時自己的心情，好像粉絲見到偶像一樣，心情十分興奮。）

　　初次跟韋太上課，已感受到她很有耐性地講解每一個製作過程及細節，以及毫無保留地將她知道的全部知識教授給每一位學員，她真是一位值得大家尊重的導師。

　　我倆自從那天開始已很投緣，在我返回荷蘭前，我倆還相約吃飯，大家盡訴心中情，無所不談，自此便成為好朋友。

　　我返回荷蘭後，第一時間使用韋太教授給我的知識製作麵包，初期由於需要練習及控制爐溫，雖然製成品自己滿意，但韋太認為還要在爐溫上下功夫，經過調整後，成品已令家人一致讚賞，使我雀躍萬分，每天都要製作麵包。

　　韋太得知我十分投入及喜愛製作麵包，所以我們每天都有聯絡，還透過網絡電話實行越洋教學。她還親自將製作過程拍成短片給我，力求仿如親自教授一樣。其後，她還收了我這個「麵包痴」為徒弟，使我更努力地學習。

　　韋太個性率直，對每一個製成品要求很高，她對每位朋友都付出一顆真心及愛心，她認為分享是一種美德。她的網站內刊出不同的食譜及影片，目的是跟大家分享，每一個食譜都詳盡地講解，而且成功率非常高，我曾試過很多網上食譜，但都失敗，這是唯一的食譜網站，能使我一次就成功，味道當然很好！

　　這次有幸獲邀寫序，作為一個朋友及粉絲當然十分開心，另外，作為徒弟，師父能將她的教學方法發揚光大，讓更多人學習，當然高興萬分。

　　最後，祝願韋太（師父）這本新著作能夠使每位讀者親身感受自家製作愛心麵包的樂趣並與家人朋友分享。

海外粉絲及徒兒

Joey

Foreword

I got to know Mrs Wai due to a chicken pie recipe. In 2009, I found Mrs Wai's website when I surfed the Internet to find recipes, I tried the recipes immediately and the result of the first trial is successful! I then became a big fan of Mrs Wai.

I have been supporting Mrs Wai silently, when I came back to Hong Kong in December last year (2011), I took the opportunity to learn to make bread from Mrs Wai. (Remember the feeling was very excited, like a fan to see the idol.)

When had class in person with Mrs Wai at the first time, she was very patient in explaining the process and details of each step, she shared all knowledge that she knew to every student without reservation, she was a mentor worth respect.

We got along well since that day. We also had dinner before I returned to the Netherlands, we became good friends since then.

I tried to make bread immediately when I returned to the Netherlands. I was satisfaction although I need to practise and control oven temperature initially. However, Mrs Wai thought that there will be improvements about adjusting oven temperature. After the adjustments, the bread gained much praise from my family, I jumped for joy and make bread every day.

When Mrs Wai learned that I am very dedicated and love making bread, we contact every day. She then teaches me overseas through the telephone network. She also makes videos of making bread to show me, just like teaching me in person. Subsequently, she accepts me, the bread-lover, as her apprentice, I then work harder to learn.

Mrs Wai is outspoken and is demanding towards every produce. She thought that sharing is a virtue and she is sincere and loving to her every friend. She shares varous recipes and videos in her website, the purpose is to share with the public.She explains each recipe in detail and the success rate is very high, I have tried many online recipes and have failed. This is the only recipe website that I was successful at the first trial, of course, the taste is good!

As a friend and a fan, I am very happy to be invited to write a foreword for my master. Also, as an apprentice, I am pleased to see that my master share her teaching methods and to enable more people to learn.

Finally, best wishes to Mrs Wai (master) and hope tha t this cookbook enables every reader to experience the fun of homemade loving bread and share with family and friends.

Overseas fan and apprentice

Joey

自序

　　我很開心今次有幸首次與橘子文化合作。起初我以為這次可以來個突破，出版個人中菜食譜，但我知道市面上已經有很多中菜食譜，要突圍而出是較困難的事，而出版社今次邀請筆者合作，希望能夠將自家製作麵包與讀者分享。

　　因為市面出售的麵包食譜，大部分都是經由專業師父或大型機器製作，要使用的工具都是較為專業，而沒有考慮一般家庭如何負擔或操作。

　　筆者利用公餘時間教授麵包製作課程，全部採用手工搓揉方法。至今教授的學生超過五百人，而且大部分學生都是新手、粉絲或在其他地方學習後仍不懂的學生，當然有小部分是海外粉絲及準備移居海外的朋友。

　　今次有幸編寫《第一次學做手工麵包！超安心》，當然要將我所學到的知識教授予大家，力求使每位讀者及粉絲百分之百了解製作麵包的過程，並能夠與家人分享這份喜悅。

　　現今市面所出售的麵包，大部分會加入添加劑及防腐劑，而我所教授自家製作的麵包，當然是採用天然的方法製作，力求讓大家吃得安心健康。

　　最後，當然要謝謝我家的三位大少爺（韋老爺、韋大少、韋二少），他們時刻協助我做試食專家且給予批評，筆者才有今日的成就及進步。

　　新讀者如對食譜內容有疑問，歡迎隨時到訪網站參觀及交流下廚心得。

韋太

Preface

I am delighted and honored to work with Wan Li Book Co Ltd the first time. At first I thought this can come to a breakthrough in publishing a cookbook with Chinese dishes, but I understand that there are numerous Chinese dish recipes in the market and it could be more difficult to stand out. The publisher invited me to share with readers homemade bread.

Since most of the bread cookbook for sale are produced by professional chef and large and professional machines and tools are used, how the family handles or operates may not be considered.

I teachs bread making co urses in spare time, all are hand-made method. The number of students is over five hundreds. Most of the students are beginners and fans or students who do not know how to make bread after taking other courses. Of course, some are overseas fans and friends plan to emigrate.

I am honored to write this cookbook, I want to share with everyone the knowledge I have learned. I hope that every reader and fan fully understand the process of making bread, and be able to share the joy with their families.

Additives and preservatives are added in most bread sold in the market today. While the homemade bread I teach, of course, is using natural methods of production considering readers' health.

Finally, I would like to thank my family of three masters (my husband and two sons). For they always help as the tasting experts and give criticism, or I will not achieve the progress.

For new readers, please feel free to visit Mrs Wai's website cooking classroom.

Mrs Wai

目錄 Contents

特色麵包

湯種麵包

健康麵包

方形麵包

不用烤的麵包

麵包入門的理論課
Basic Theory of Bread

如何在家做出鬆軟的麵包？

要做出鬆軟的麵包，就好像要過五關斬六將一樣。

除了要在做麵包初期控制麵糰的筋性及水分之外，入烤箱的溫度及如何保存製成品的品質，都是一門重要的學問。

很多人認為，做麵包最難的一關就是發酵。發酵的時間長短，會隨着一年四季的天氣變化而受到影響。如果大家明白當中受影響的因素，就可以自行判斷而作出調整，這樣就更容易做出鬆軟的麵包。

一般家庭式自製麵包，大多數會用手去搓揉。做出來的麵包筋性會比較用麵包機的強及富彈力。而用大馬力的攪拌機，雖然可以做出相同效果，但並不是每個家庭都能負擔。

鑒於家庭式及自家製的關係，這本書著重以家庭式的方法及份量去製作，務求令每位讀者容易明白及操控。

請大家切勿心急開始動手做，懇請花少許時間細心閱讀以下內容，以及可能在製作過程中遇到的問題，開始時按照每個步驟檢查或測試，便能夠事半功倍。

How to male a soft bread at home?

To make a soft bread, we need to understand the basic techniques and have many trials.

In addition to control the gluten and moisture of making bread dough in the early stages, it is also essential to handle oven temperature and store and keep the quality of bread.

Many people think that fermentation is the most difficult when making bread. The duration of the fermentation varies according to weather throughout the year. We could make own adjustments if we understand the factors, so it is easy to make a soft bread.

Many homemade bread are kneaded by hands, the gluten of bread will be stronger and more elastic compared with those made by machine. The same effect could be achieved by using a high-powered blender but not every household can afford.

In view of homemade basis, this cookbook emphasizes on the method and amount of family-style production so that each reader is easy to understand and control.

Do not hurry to start, please spend some time to read the following carefully. There are problems may be encountered in the production process. It will be helpful to inspect or test in accordance with each step.

製作麵包基本原料

五項主要原料，是麵包製作必需成分

- ・高筋麵粉、酵母、鹽、砂糖、水等為主要原料。
- ・再配搭其他原料，產生化學作用，造成不同效果。
- ・其他原料包括：雞蛋、奶粉等。

開始搓揉時，由於每天的溫度及濕度都不同，初學者要留意以下事項

酵母測試

買回來的酵母，開封後一定要存放在冰箱的冷藏室。若放在室溫或廚房等濕氣重或太暖的地方，酵母較容易與室內的暖空氣接觸而釋放出二氧化碳，等到使用時，酵母的壽命差不多已盡，不能在製作過程中發揮效用，因而使麵糰無法膨脹，影響製作過程及不能繼續進行下去。

本書所用的酵母是即溶酵母（或稱速效酵母），即溶酵母無需預先浸泡，可以立即搓揉，使用方便。

一般超市購買到的乾酵母由於顆粒較粗，建議使用前浸泡5-10分鐘，確保酵母有活躍力。有人喜歡用新鮮酵母，因它製作出來的麵包風味較佳，但保存期有限也不容易買到。所以，本書建議用即溶酵母或乾酵母。

不同的天氣，加入不同的水分

水分要按照每天不同的溫度及濕度自行調整，但這樣說法對於初學者而言比較困難，如果能夠明白以下的要點，就比較容易操作。

麵糰要有適當的溫度及濕度才容易進行發酵，最理想的室內溫度約28-30℃，相對濕度為80-85%。

因此，如果當天氣溫較低且相對濕度較低（即天氣寒冷及乾燥），室內水分容易流失，麵糰缺水的可能性較大，所以開始計算水的份量時，便要較平日的份量略為增加5-10毫升水，以補回流失的水分。麵糰太乾，搓揉時會比較吃力，如果開始時能夠提供足夠的水分，就容易起步及達至預期效果。

相反，室內溫度及相對濕度較高（即天氣和暖及潮濕），室內水分不容易流失，麵糰流失水分較少，建議開始時不用增加水分，待中途有需要時再適量地調整。

麵糰柔軟度測試

本書教授的搓揉方法與一般食譜的搓揉方法不同。一般方法建議最後才將奶油（或酥油）加入搓揉，但本食譜建議一開始就全部放進搓揉，其原理與麵包機製作大同小異。只要大家用心搓揉及學懂如何查看麵糰的水分及柔軟度是否達標，即使是初學者，也能揉出富有彈性且鬆軟的麵包。

本食譜的搓揉方法，對初學者來說，可能覺得麵糰偏濕也難於操控，但如果開始時沒有加入奶油，搓揉時又會覺得太乾，麵糰水分不足，搓揉起來較困難，要搓揉至起筋性更難。若水分適當，麵糰摔打在桌面時會較容易，麵糰長度約有8-10吋，即麵糰柔軟度足夠，在搓揉時麵糰沒有大量流失水分，這樣，待完成第一次發酵後，麵糰會較為鬆軟。

建議初學者搓揉時間約15分鐘，因為若超過這時間，麵糰水分會不斷地流失，若過於強求如其他食譜所說的搓揉至有彈性可拉出薄膜，對於初學者來講會較困難。時間到卻不能搓揉至有薄膜時，便要停止動作。待下回再將搓揉的力度和速度加快，才容易達至最佳的效果。

入烤箱溫度調整（爐溫測試）

每個家用烤箱的溫度都有差異，就算是相同品牌，爐溫都不一。

即使麵糰能夠在初期保持濕潤，如果爐溫持續過高，幾分鐘的時間便會將麵包內的水分抽乾，因而使製成品欠缺水分及彈性。

若要製成品達至理想效果，請務必在初次烤製麵包時進行1-2次爐溫測試。

要進行爐溫測試，請將烤箱的溫度調整至180℃的一般要求溫度，並預熱10-15分鐘。一般麵包烤製時間約15分鐘，吐司要烤30-35分鐘。

請將完成最後發酵的麵包放入已預熱的烤箱，烤箱時間撥至30分鐘（電子烤箱不用這樣調整），另外再用計時器計時7分鐘，7分鐘後，要觀察麵包表面顏色是否上色或太深。若顏色偏深色，爐溫有可能比正常溫度高，要即時將爐溫調減10-20℃。如果使用沒有熱風對流的烤箱，靠近烤箱內側的一邊，麵包會較容易上色，這時可以快速將烤盤前後位置調動。相反，顏色較淺，有可能爐溫偏低，要即時調高10-20℃繼續測試及完成烘烤。

所以，要即時補救，便要用剩下的8分鐘時間去補救。但下次再用這烤箱時，就要調整好後期已修正過的爐溫去烤製。

如果發現麵包仍未上色或烤熟，烤製時間最多補回5-10分鐘。

記得每多烤一分鐘，麵包的水分就容易散失使麵包內部變得粗糙及乾燥，缺乏彈性及水分。

烤箱時間調整至30分鐘，原因是烤箱的電熱管會在剩下的10分鐘左右，隨著時間倒退而開始降低溫度，為確保爐溫平均，所以建議另外用計時器計時。

出爐的麵包

要知道麵包是否已烤熟，在指定時間到時，先看看麵包的底部是否乾爽且呈現金黃色，如果是即表示麵包已熟。出爐後應先將麵包放在網架上冷卻一會才食用，以免麵包組織黏在一起。

如何保存製成品

要有適當方法才可以保持麵包濕潤且延長麵包老化期。

自家製麵包由於沒有加入任何添加劑，較一般市售麵包容易變得乾硬。所以，若出爐的麵包不是馬上食用，放涼後，一定要先放入保鮮盒或密封袋內，即時鎖住麵包水分，才放入冰箱冷藏。這樣，麵包保鮮期可以有3-4天，加入湯種的麵包則可保存一星期。

食用前，只需放入微波爐加熱30秒鐘，如果時間過長，會抽乾麵包的水分。若無微波爐，可以用蒸的方法加熱，但千萬不要用烤箱。

如果大家能夠在開始學習時細讀以上各項要點，必能達到理想效果。

Basic Ingredients for Making Bread

Five essential components for making bread

- High-gluten flour, yeast, salt, sugar, water, etc. as the basic ingredients.
- Then add other ingredients to create chemical reaction resulting in different effects.
- Other ingredients include eggs, milk, etc.

> **When begin kneading, beginners should note the following matters due to different daily temperature and humidity**

Yeast test

Yeast must be stored in refrigerator after opening. If yeast is stored at room temperature or humid or warm place in the kitchen, carbon dioxide will be released when contact with heated air in the room. Yeast cannot be effective in the production process as a result, the size of the dough cannot expand, the production process cannot be continued.

Instant yeast (or quick yeast) is used in this cookbook. No need to soak in advance, instant yeast can be kneaded instantly and conveniently.

Dried yeast bought in supermarket are having coarse particles, it is recommended to soak for 5-10 minutes before use to ensure that the yeast has active capacity. Some people like to use fresh yeast because it makes bread flavor better. However, the expiry period of fresh yeast is limited and not readily available. Therefore, instant yeast or dried yeast is recommended by this cookbook.

Add different amount of water according to different weather

The amount of water should be adjusted by daily temperature and humidity. But this saying may be difficult for beginners. It will be easier to operate if we understand the following points.

The dough will be easier to ferment when having appropriate temperature and humidity. The ideal room temperature is about 28-30℃,

relative humidity is 80-85%.

Therefore, if the temperature and relative humidity are low (ie, the weather is cold and dry), moisture is easy to lose and the dough will be lack of water more easily. So when starting to calculate the amount of water, 5-10ml of water should be added to compensate for the loss of moisture. It is more difficult to knead if the dough is too dry. It is recommended not to add water at the beginning but add and adjust when needed.

Testing the softness of the dough

The method of kneading is different from general recipes in this cookbook. The general approach suggests to add butter (or shortening) finally. But this cookbook suggests to put in all ingredients at the beginning, the principle is more or less like using machine. As long as we knead the dough by heart and learn how to test the moisture and softness of the dough, elastic and soft bread could be made even by beginners.

Beginners may feel that the kneading method in this cookbook is difficult to handle as the dough is relativle wet and difficult to manipulate. However, if butter is not added at the beginning, the dough will be too dry. If the dough is not moist enough, it is more difficult to knead until gluten forms. If the amount of water is correct, it is easier to beat the dough onto the table and the length of the dough is about 8-10 inches. That is softness of the dough is appropriate, the moist of the dough did not lost seriously during kneading. In this way, the dough will be softer after first fermentation.

It is recommended that beginners knead for about 15 minutes, the moisture of the dough will be lost if beyond this time limit. It will be more difficult for beginners to knead until the dough is elastic and a membrane can be pulled as mentioned in other cookbooks. We have to stop when time is up even if a membrane cannot be pulled. Then we need to rub the dough intensely and improve the speed next time in order to achieve the best result.

Adjusting oven temperature (oven temperature test)

The temperature of each household oven varies, the temperature is different even of the same brand.

Even though the moisture of the dough can be kept at an early stage, the moisture of the bread could be lost within a few minutes if oven temperature is too high, hence the bread will be lack of moisture and elasticity. So 1-2 times initial oven temperature tests before baking bread is essential to achieve the desired effect.

When testing oven temperature, adjust the oven to 180℃ according to the general requirements and preheat for 10-15 minutes. General baking time for bread is about 15 minutes, square bread need to bake for 30-35 minutes.

Put the bread into the preheated oven after final fermentation, turn the oven timer to 30 minutes (need not adjust like this for electronic oven). Set another timer to 7 minutes, observe the surface color of the bread after 7 minutes and check if the color is too dark. If the color is too dark, the oven temperature may be higher than normal temperature, reduce the oven temperature by 10-20℃ immediately. If you are using an oven without hot air convection, the color of the bread will be darker near the inside of the oven, then you can reverse the baking tray quickly. On the contrary, if the color is lighter, it is likely that the oven temperature is relatively low, increase the oven temperature by 10-20℃ immediately and continue to test and complete the baking.

Therefore, immediate remedy is necessary and we use the remaining 8 minutes to remedy the situation. It is necessary to adjust oven temperature at the very beginning when baking next time.

If the bread is not yet painted or baked till done, bake for additional 5-10 minutes at most.Bear in mind that every extra minute to bake the bread can make the moisture loss and hence make the inner of the bread rough and dry, lack of elasticity and moisture.

The oven timer is adjusted to 30 minutes because the heater of the oven will reduce the temperature in the remaining 10 minutes. In order to ensure that oven temperature is average, it is recommended to use another timer to count.

Bread finish baking

To check if the bread is done after specified time, check if the bottom of the bread is dry and golden brown. If yes, it means that the bread is cooked. After taking out the bread from the oven, put onto a rack for cooling for a while before eating, otherwise the tissues of the bread will be sticked together.

How to store bread

Appropriate method for storing bread can keep the bread moist and it could be kept for longer period.

Since no additives is added for homemade bread, it will become dry and hard easily compared with those sold in the market.

Therefore, if the baked bread is not eaten immediately, leave to cool and put into a box or a bag, lock the moisture of the bread instantly before refrigerate. In this way, the bread could be kept for 3-4 days. For those bread with Utane starter added, it can be stored for one week.

When eating, reheat the bread in microwave oven for 30 seconds, do not reheat for too long or the moisture of the bread will be lost.

If there is no microwave oven, reheat by steaming, do not use the oven.

If we can bear in mind the above points before start learning, the desired effect will be achieved.

麵包基本材料分析

麵粉

是由小麥磨製而成。

小麥一般分為「軟麥」和「硬麥」兩種。軟麥磨出來的叫低筋麵粉，硬麥磨出來的叫高筋麵粉。其實一般麵粉分為三大類：低筋麵粉、中筋麵粉及高筋麵粉。

一般麵包製作會選用高筋麵粉。麵粉的筋度高低取決於小麥粉中的蛋白質含量。（海外的朋友可以按照包裝上的蛋白質成分辨別哪些是高筋麵粉。）

低筋麵粉

蛋白質含量約 6.5%-9%，適合用於製作各式蛋糕、餅乾等鬆軟糕點。如果沒有低筋麵粉，也可用中筋麵粉和20%的玉米粉取代，玉米粉可降低麵粉的筋性。

中筋麵粉

蛋白質含量約9%-11.5%，一般超級市場售賣的普通麵粉 (Plain flour / All-purpose Flour) 是中筋麵粉。適用於製作饅頭、包子、水餃皮等。

高筋麵粉

蛋白質含量約11.5% - 14%，加水搓揉會出現彈性，適合用來製作麵包。

酵母

酵母分為乾酵母和新鮮酵母。作用是將碳水化合物轉化成二氧化碳及酒精。（若麵糰過度發酵，麵包會有酸味。）

鹽

鹽除了調節麵包的味道，還可以用來調整發酵時間，減緩酵母的發酵速度。亦能改變麵筋的性質，增加麵糰的吸水能力，使麵筋膨脹而不致斷裂。

砂糖

砂糖有助增加酵母的活躍能力，促進麵糰發酵作用，有助改進麵包的色澤、鬆軟度及保存期，但糖分過高或超過5%都會抑制麵包的發酵。

水

水可使各種乾性原料充分混合。加入適量的水，可控制麵糰的稠度、柔軟度及黏性。而用溫水則有助酵母發酵。

奶粉

可提供奶香味和增加色澤，也具有增強麵糰筋度、延遲老化、使組織細緻等作用。

Basic Ingredients of Bread

Flour

Fllour is made by grinding whole wheat.

"Soft wheat" and "hard wheat" is the two main categories of wheat. Low gluten flour is ground from soft wheat while high gluten flour is ground from hard wheat. In fact, flour is divided into three categories: low gluten flour, all-purpose flour and high gluten flour.

High gluten flour is usually used in bread making (referred to as flour). The degree of gluten depends on the protein content in wheat flour. (Readers living abroad could identify high gluten flour according to the protein content on the package.)

Low gluten flour

Also known as cake flour, the protein content is 6.5% -9%, suitable for making all kinds of cakes, cookies and other soft cakes. If there is no low gluten flour, all purpose flour and 20% of cornstarch could be used instead. Cornstarch can reduce the gluten of the flour.

All-purpose flour

The protein content is 9% - 11.5%. Flour (plain flour) sold in supermarket is all-purpose flour. Suitable for making steamed buns, Chinese buns and dumpling wrappers.

High gluten flour

Also known as bread flour, protein content is 11.5% - 14%. The dough will be elastic when the flour is knead after adding water. Suitable for making bread.

Yeast

Yeast could be divided into dried yeast and fresh yeast. Its role is to transform carbohydrates into carbon dioxide and alcohol. (The bread will taste sour if the dough is over-fermented.)

Salt

Besides adjusting taste of bread, it can be used to adjust the time of fermentation, it could slow down the speed of yeast fermentation.Salt can also change the nature of the gluten by increasing the absorbent capacity of the dough, expand the gluten to avoid breaking.

Sugar

Sugar can help to enhance the activity of yeast. It also promotes the dough fermentation and improves the color, softness and storage of bread. But if too much sugar, say more than 5%, is added, it will restrain the fermentation of bread.

Water

Water can be fully mixed with dry materials. By adding an appropriate amount of water, we can control the consistency, softness and stickiness of the dough. Warm water helps the fermentation of yeast.

Milk powder

It provides fragrance of milk flavor, enhances the color and degree of elasticity of dough, delay aging and make the dough more meticulous.

配合麵糰發酵的元素

- 室內溫度：30℃，相對濕度 80% - 85%
- 麵糰溫度：28℃
- 第一次發酵時間：35-45 分鐘
- 取出、排氣、分割及鬆馳：15-20分鐘
- 造型
- 最後發酵時間：30-45分鐘

若操作時間過久，麵糰內部會因發酵過度而產生過多的酸，導致麵糰老化，影響麵糰性質。因此時間的控制是整個製作過程中最重要的工作。

一般市面售賣的麵包都會加入添加劑，以增加麵包的柔軟度。麵包改良劑可增加麵糰中的麵筋，維持搓揉後麵糰的網狀結構，增加麵包的體積，使口感更佳。但本書著重自家製、無添加及健康，所以並不建議加入任何麵包改良劑或添加劑。

鹹麵包含的糖分較甜麵包低，烘烤時間可以多1-2分鐘。相反，甜麵包含的糖分較高，進烤箱後要留心觀察顏色的變化。

如何分辨麵糰是否搓揉恰當？

麵糰用手拉時有彈性且容易拉出一小塊透明的薄膜。麵包出爐時，表面有光澤且豐滿，彈性好，體積脹大，咬下去有口感，保存期也相對較長。

相反，麵糰搓揉不足，用手拉開麵糰時欠缺彈性，容易斷裂，內部組織粗糙。

最後發酵注意事項

要使麵糰達到理想發酵，就要配合發酵必要的元素，室內的溫度，應維持約30℃。

溫度過高會使麵糰內外發酵的速度不同，烤出來的麵包也會失去原有的風味。

嚴重時，會破壞麵包的形狀，也會使體積縮小而形狀扁平。

若遇氣溫較低時，在最後發酵時室溫太低，則發酵速度緩慢，發酵時間延長，麵糰的性質差異愈大，造成麵包的品質不良。因此必須特別注意發酵的溫度和時間操控。

最後，發酵的相對濕度應維持在80%-85%。濕度太低，麵糰表皮容易乾，做出來的麵包會欠缺彈性。若濕度過高，麵包的品質受害更大，內部組織及外形會受到極嚴重的破壞。

Elements enhancing the fermentation of dough

- Rome temperature: 30°C, relative humidity 80% - 85%
- Dough temperature: 28°C
- First fermentation time: 35-45 minutes
- Take out, release gas, separate and rest: 15-20 minutes
- Shape
- Final fermentation time: 30-45 minutes

If the operation takes too long, the dough will be over fermented, too much acid will be generated internally, and the dough will be aging and properties of the dough will be affected. Time control is the most important task during the entire process of production.

Bread additives will be added to most of the bread sold in the market to increase the softness of the bread.

Flour modifier can increase the gluten of the dough, consolidate the network structure of the dough after kneading, increase the volume of bread, and hence make the bread tastes better. Since this cookbook emphasizes homemade, no additives and health-oriented, so no flour modifier or bread additives are recommended.

Salty bread contains less sugar, the baking time could be 1-2 minutes more. In contrast, sweet bread contains more sugar, we should observe the color changes carefully.

How to differentiate if the dough is well kneaded?

Dough is elastic when stretched by hands and it is easy to pull out a small piece of transparent membrane. When taken out from the oven, the surface of the bread is shiny and plump, elastic, full body, the bread could be kept for longer.

On the contrary, if the kneading time of dough is not enough, it will be inelastic, easily broken, the internal structure is rough.

Note for Final Fermentation

To make a good dough, there should be necessary elements for fermentation, the room temperature should be maintained around 30°C.

If the temperature is too high, it will affect the speed of fermentation inside and outside, the flavor will be affected.

The shape of bread will be deteroiated, the size will be small and the shape will be flat.

In case of low temperatures, the room temperature for final fermentation is too low, extending the fermentation time. The differences in the nature of the dough the greater, resulting in poor quality of the bread. Fermentation temperature and time control must be monitored closely.

The relative humidity should be maintained between 80%-85% for final fermentation. If humidity is too low, the surface of the dough will be dried easily and the bread made will be lack of elasticity. If the humidity is too high, the quality of bread will be affected more severly, the internal structure and shape of the bread will be damaged very seriously.

基本用具
Basic Utensils

1. 鋁箔紙
   aluminium foil

2. 保鮮膜
   plastic wrap

3. 桿麵棍
   rolling pin

4. 12吋長膠尺
   12" ruler

5. 刷子
   brush

6. 450克有蓋吐司模
   450g square mold with lid

7. 噴水器
   spray bottle

8. 電子秤
   electric scale

9. 鋼盆
   steel basin

10. 塑膠刮板
    plastic scraper

11. 毛巾
    towel

12. 計時器
    timer

13. 圓形鋁箔紙
    round aluminium foil

14. 量匙
    measuring spoon

15. 刀
    cutter

16. 6吋戚風蛋糕模
    6" chiffan cake mold

17. 6吋方形模
    6" square mold

18. 四吋圓形模
    4" round mold

19. 麵包放涼架
    rack for cooling bread

20. 烤盤
    baking tray

21. 麵包夾
    bread clip

白醬雞肉餡 8個份量（8 servings）
Chicken filling in white sauce

材料 Ingredients

雞肉　150克/g
chicken

洋葱（切丁）　1/2個
onion (diced)

磨菇（切片）　4粒
button mushrooms (sliced)

奶油（起鍋用）　1茶匙/tsp
butter (for heating the wok)

雞肉醃料 Marinade for Chicken

淡醬油　2茶匙/tsps
light soy sauce

紹興酒　1茶匙/tsp
Shaoxing wine

玉米粉　1茶匙/tsp
cornstarch

麻油及胡椒粉　各少許
sesame oil and a pinch of pepper

白醬材料 Sauce

奶油　20克/g
butter

麵粉　1湯匙/tbsp
flour

水　100毫升/ml
water

牛奶　30克/g
milk

雞粉　1茶匙/tsp
chicken powder

鹽　1/8茶匙/tsp
salt

糖　1/8茶匙/tsp
sugar

胡椒粉　少許
some pepper

作法 Method

1. 雞肉切丁，加入醃料醃15分鐘，洋葱切細丁備用。

2. 將鍋子燒熱，加入1茶匙奶油，將洋葱炒香後，加入雞丁炒至半熟，然後加入其他材料略炒後盛起備用。

3. 將鍋子另燒熱，加入奶油20克煮溶後，加入麵粉，快速將麵粉炒至沒有粉粒，將已炒好的餡料加入拌勻。然後將其他白醬材料加入，以慢火將醬汁煮至濃稠，關掉爐火，放涼備用。

1. Cut chicken into cubes, marinate for 15 minutes. Finely dice onion and leave for later use.

2. Heat the wok, add 1 tsp of butter, sauté onion until fragrant, add diced chicken and stir-fry until half-cooked, then add other ingredients and stir-fry, dish up and set aside.

3. Heat another wok, add 20g butter and cook until melts, add flour and stir-fry quickly until smooth, add all the stir-fried ingredients and stir well. Add sauce, simmer and cook until sauce thickens, turn off heat, leave to cool for later use.

叉燒包餡
Filling for barbecued pork bun

8個份量，每個約20克
（8 servings, each about 20g）

材料 Ingredients

叉燒（切粒） 120克/g
barbecued pork (diced)

乾葱末 2茶匙/tsps
chopped shallots

醬汁材料 Sauce

蠔油 1湯匙/tbsp
oyster sauce

陳年醬酒 1茶匙/tsp
dark soy sauce

淡醬酒 2茶匙/tsps
light soy sauce

砂糖 1/2茶匙/tsp
sugar

麻油 1/2茶匙/tsp
sesame oil

雞粉 1/4茶匙/tsp
chicken powder

鹽 1/8茶匙/tsp
salt

胡椒粉 少許
some pepper

水 60毫升/ml（4湯匙/tbsps）
water

麵包表面 Syrup for bun surface

砂糖 1茶匙/tsp
sugar

熱 1/2湯匙/tbsp
hot water

芡汁 Thickening

玉米粉 1茶匙/tsp
cornstarch

清水 2湯匙/tbsps
water

起士培根餡
Cheese and bacon filling

材料 Ingredients

培根（切粒） 2條/strips
bacon (diced)

馬自瑞拉起士 適量
some mozzarella

做法 Method

將所有材料拌勻。
Mix all ingredients well.

作法 Method

1. 將鍋子燒熱，加入油1/2湯匙，將乾葱末加入爆香，然後將醬汁材料加入煮滾。

2. 將芡汁拌勻，加入煮至滾起，加入叉燒粒拌勻。放涼後，放入冰箱冷藏一會備用。

1. Heat wok with 1/2 tbsp of oil, sauté chopped shallots until fragrant, then add sauce and bring to a boil.

2. Mix and add thickening and bring to a boil, add chopped barbecued pork and mix well. Leave to cool and refrigerate for a while.

香草玉米牛肉餡 8個份量
Corn and beef filling with herbs
（8 servings）

材料 Ingredients

鹹牛肉　150克/g
corned beef

玉米粒　50克/g
corn kernels

香草　1/2茶匙/tsp
herbs

細鹽　1/8茶匙/tsp
fine salt

蛋黃醬或沙拉醬　適量
some mayonnaise or
salad dressing

做法 Method

將所有材料拌勻。
Mix all ingredients well.

榛子醬
Hazelnut paste

（4個麵糰）榛子醬　適量
(4 doughs) some hazelnut paste

椰香餡
Coconut filling

1份麵糰，6吋x12吋，可切出8-10個小椰香卷（2個）
（1 dough, 6" x 12", can be cut into 8-10 mini coconut buns (two)）

奶油蛋糕模 butter cake mold

2個，9吋x5吋（2, 9" x 5"）
4個，3吋x6吋（4, 3" x 6"）

材料 Ingredients

椰絲　25克/g
shredded coconut

砂糖　25克/g
sugar

奶油（融液）　25克/g
butter (melted solution)

做法 Method

將所有材料拌勻。
Mix all ingredients well.

麵糰餡料 Dough filling

將一份麵糰切成2份，放入
兩個奶油蛋糕模內。
cut one dough into 2
portions, place into 2
butter cake molds.

香芋番薯餡
Taro and sweet potato
square bread

材料 Ingredients

2個半磅方形麵包
(2 half-pound square bread)

芋頭（切細絲）　100克/g
taro (finely shredded)

糖粉　25克/g
icing sugar

做法 Method

蒸熟，加入糖粉拌勻。
Steam taro till done, add icing sugar
and mix well.

鮪魚蛋白餡
Tuna and egg white bread

材料 Ingredients

鮪魚（1小罐） 180克/g
tuna (1 small can)

玉米粒 40克/g
corn kernels

細鹽 少許
some fine salt

香草 少許
some herbs

蛋黃醬或沙拉醬 適量
some mayonnaise or
salad dressing

做法 Method

將所有材料拌勻。
Mix all ingredients
well.

紫心蕃薯餡
8個份量
（8 servings）
Purple sweet potatoe bread

材料 Ingredients

番薯 100克/g
sweet potato

做法 Method

將番薯蒸熟去皮搗泥。
Steam sweet potato until
done, peel and smash.

火腿葱餡
8個份量
（8 servings）
Ham and spring onion filling

材料 Ingredients

火腿 50克/g
ham

葱 10克/g
spring onion

雞粉及麻油 各少許
some chicken powder and sesame oil

做法 Method

火腿和葱分別切粒，放入盆內，加入雞粉及麻油
拌勻。
Dice ham and onion respectively and put into a
basin, add chicken powder and sesame oil, mix
well.

香蒜包-蒜蓉醬
Garlic bread - garlic paste

材料 Ingredients

奶油　1湯匙/tbsp
butter

蒜泥　2茶匙/tsps
chopped garlic

香草　1/2茶匙/tsp
herbs

細鹽　1/8茶匙/tsp
fine salt

做法 Method

將所有材料拌勻。
Mix all ingredients well.

雞尾麵包內餡
Cocktail bun filling

材料 Ingredients

奶油（硬）　70克/g
butter (hard)

麵粉　20克/g
flour

奶粉　20克/g
milk powder

糖　20克/g
sugar

椰絲　10克/g
shredded coconut

做法 Method

所有材料放入盆內，並將奶油切
小粒，然後慢慢揉勻成糰狀。
Add all ingredients into a basin,
finely dice butter, and then slowly
rub evenly into a dough.

照燒雞腿
Japanese Teriyaki Chicken Bun

材料 Ingredients

雞肉　3大片
chicken fillets

醃料 Marinade

日式照燒醬　3湯匙/tbsps
Japanese Teriyaki sauce

味醂　1湯匙/tbsp
mirin

玉米粉　1/2茶匙/tsp
cornstarch

雞粉　1/2茶匙/tsp
chicken powder

砂糖　1茶匙/tsp
sugar

細鹽　1/4茶匙/tsp
fine salt

胡椒粉、麻油　各少許
some pepper and
sesame oil

做法 Method

1. 雞肉洗淨去皮，瀝乾水。每
份雞肉切成2片。（厚的肉
可以切薄）

2. 將所有醃料依序加入拌勻，
醃30分鐘。

3. 將鍋子燒熱，加入2湯匙
油，將雞肉煎至兩面金黃
色，將剩下的醃料加入煮
滾即可。

1. Rinse chicken fillets and
skin, drain. Cut each piece
of chicken fillet into halves.
(Cut into thin slices if the
fillets are too thick.)

2. Add all marinade in order,
mix well and marinate for
30 minutes.

3. Heat wok with 2 tbsps of
oil, add chicken fillets and
pan-fry until both sides
are golden brown, add the
remaining marinade and
bring to a boil. Serve.

豬肉漢堡排
Pork filling

4個份量
（4 servings）

材料 Ingredients

豬絞肉　150克/g
minced pork

洋葱（切細粒）　1/4個
onion (finely diced)

麵包粉　適量
some bread crumbs

醃料 Marinade

淡醬油和清水　各1湯匙/tbsp each
light soy sauce and water

玉米粉　1/2茶匙/tsp
cornstarch

香草末　1/2茶匙/tsp
chopped herbs

雞粉及砂糖各　1/4茶匙/tsp each
chicken powder and sugar

黑胡椒　1/4茶匙/tsp
ground black pepper

胡椒粉及麻油各少許
some pepper and sesame oil

做法 Method

1. 將醃料加入豬絞肉後，大力攪拌至有黏性，然後加入洋葱碎粒拌勻，放入冰箱冷藏30分鐘。

2. 將豬絞肉平均分成4份，搓整成圓形後壓平，在每塊底部沾上麵包粉，備用。

3. 將鍋子燒熱加入油2湯匙，將豬肉排，略壓成圓扁形，兩面煎至金黃色，煎熟即可。

1. Add marinade into minced pork and stir vigorously, add chopped onion and mix well, refrigerate for 30 minutes.

2. Divide minced pork into 4 portions, knead into round shape and flatten, coat each piece with bread crumbs, leave for later use.

3. Heat wok with 2 tbsps of oil, add pork fillets and press slightly into round shape, pan-fry both sides until golden brown and done. Serve.

日式咖哩牛肉餡
Japanese curry beef filling

8個份量
（8 servings）

材料 Ingredients

洋葱（切粒）　1/2個
onion (diced)

牛絞肉　200克/g
minced beef

芡汁 Thickening

麵粉　2湯匙/tbsps
flour

水（拌勻）　90毫升/ml
water (mix well)

調味料 Seasonings

日式咖哩醬　1湯匙/tbsp
Japanese curry paste

咖哩粉　1茶匙/tsp
curry powder

砂糖　1茶匙/tsp
sugar

紹興酒　1茶匙/tsp
Shaoxing wine

鹽　1/4茶匙/tsp
salt

做法 Method

1. 將鍋子燒熱加入油2湯匙，將洋葱放入炒軟，然後將牛肉加入後炒熟。

2. 加入調味料再炒勻，將芡汁拌勻後，慢慢加入並炒至濃稠後關火，放涼備用。

1. Heat 2 tbsps of oil in wok, add onion and stir-fry until soft, then add beef and stir-fry until done.

2. Add seasonings and stir well, then add thickening and stir well slowly until thickens, turn off heat and leave to cool. Set aside for later use.

墨西哥麵包內餡
Mexican filling

8個份量
（8 servings）

材料 Ingredients

軟奶油　80克/g
soft butter

麵粉　30克/g
flour

砂糖　25克/g
sugar

做法 Method

奶油及糖用手動打蛋器打勻，將麵粉加入拌勻備用。

小提醒：麵包表面不用刷蛋液，防止餡料下滑。

Beat butter and sugar with a mixer, add flour and mix well. Set aside for later use.

Note：No need to brush egg mixture onto the surface, otherwise the filling will collapse easily.

北海道麵包表面
Hokkaido topping

材料 Ingredients

已融化奶油　50克/g
melted butter

麵粉　50克/g
flour

糖粉　45克/g
icing sugar

雞蛋　1顆
egg

做法 Method

將以上材料依序加入盆內拌勻即可。（可以加入少許芝麻裝飾）

Add the above ingredients in order into a basin and mix well. (sesame seeds can be added as decoration)

南瓜泥
Pumpkin pureé

材料 Ingredients

南瓜　100克/g
pumpkin

做法 Method

南瓜去皮，切薄片，蒸10-15分鐘至熟，壓成泥。

Peel pumpkin, cut into thin slices and steam for 10-15 minutes until done, mash into pureé.

菠蘿麵包皮
Cheese and bacon filling

8個份量
（8 servings）

材料 Ingredients

麵粉　50克/g
flour

酥油　29克/g
shortening

砂糖　23克/g
sugar

雞蛋　2克/g（約1茶匙/tsp）
egg

牛奶　2克/g
milk

奶粉　1茶匙/tsp
milk powder

食用臭粉　1/8茶匙/tsp
edible smelly powder

蘇打粉　1/8茶匙/tsp
baking soda

檸檬黃色素　少許
some tartrazine

做法 Method

食用臭粉、砂糖、蘇打粉預先拌勻，然後加入其餘材料拌勻即成。

Mix edible smelly powder, sugar and baking soda well, then add the remaining ingredients and mix well. Serve.

卡士達起士餡（墨西哥卡士達餡）
Custard and cream filling

8個份量
（8 servings）

材料 Ingredients

牛奶　200毫升/ml
milk

即溶卡士達粉　65克/g
instant custard powder

做法 Method

將溫牛奶加入即融卡士達粉內，然後拌勻至濃稠狀。

Add warm milk into instant custard powder, then mix well until thickens.

巧克力卡士達麵包表面
Chocolate custard topping

8個份量
（8 servings）

材料 Ingredients

奶油　50克/g
butter

糖粉　45克/g
icing sugar

麵粉　50克/g
flour

雞蛋　1顆
egg

可可粉　1茶匙/tsp
cocoa powder

白芝麻　1茶匙/tsp
white sesame seeds

做法 Method

將以上材料順序加入盆內拌勻即可。

Mix the above ingredients in order into a basin and mix well.

巧克力卡士達餡
Cheese and bacon filling

8個份量
（8 servings）

材料 Ingredients

牛奶　200毫升/ml
milk

即溶卡士達粉　65克/g
instant custard powder

巧克力豆　40克/g
chocolate chips

可可粉　10克/g
cocoa powder

做法 Method

1. 將溫牛奶加入即溶卡士達粉內，然後用手動打蛋器將麵糊攪拌至濃稠。

2. 加入可可粉拌勻，最後加入巧克力豆即可。

1. Add warm milk into instant custard powder, beat with a mixer until the batter is thickened.

2. Add cocoa powder and mix well, finally add chocolate chips. Serve.

奶黃餡
Custard filling
 15粒份量，每粒18克
（15balls, 18g each）

材料 Ingredients

砂糖　50克/g
sugar

卡士達粉　30克/g
custard powder

奶粉　20克/g
milk powder

椰奶　65毫升/ml
coconut milk

淡奶　40毫升/ml
evaporated milk

雞蛋　1顆
egg

已融化奶油　15克/g
melted butter

做法 Method

1. 將所有材料（除了已融化奶油）加入盆內拌勻至沒有粉粒，然後加入已融化奶油拌勻，放入已刷油的蒸盤上，大火蒸約7-8分鐘至熟。

2. 將已蒸好的奶黃餡放涼一會，然後放入冰箱急速冷凍3-5分鐘。

3. 將奶黃餡放在桌上，用手揉至光滑。然後將餡料平均分成等份備用。

小提醒：假如不用奶粉，可改用相同份量的卡士達粉代替，或改用玉米粉、麵粉各10克代替。

1. Add all Ingredients (except butter solution) into a basin and mix well till smooth, then add butter solution and mix well. Pour onto a greased steaming plate and steam over high heat for 7-8 minutes until done.

2. Leave the steamed dough to cool for a while, and then refrigerate for 3-5 minutes.

3. Put the dough on the table, knead into a smooth dough. Divide the filling evenly and set aside.

Note：Milk powder could be replaced by the same amount of custard powder instead, or 10g cornstarch and 10g flour.

肉桂葡萄方形麵包餡
Cinnamon and raisin filling

2個半磅方形麵包份量
（two half-pound square bread）

材料 Ingredients

葡萄乾　30克/g
raisins

蘭姆酒　1茶匙/tsp
rum

肉桂粉　適量
some cinnamon powder

做法 Method

葡萄乾用熱水略沖表面，將水分瀝乾加入蘭姆酒拌勻，浸泡一晚。

Rinse the surface of raisins with hot water briefly, drain, add rum and mix well, leave overnight.

南瓜卡士達餡
Pumpkin custard filling

8個份量
（8 servings）

材料 Ingredients

卡士達粉（免煮）　66克/g
custard powder (no need to cook)

牛奶　140毫升/ml
milk

南瓜泥　80克/g
pumpkin pureé

做法 Method

將牛奶煮滾，加入卡士達粉內，用手動打蛋器攪拌至沒有粉粒，然後加入南瓜泥拌勻，放涼備用。

Bring milk to a boil, add into custard powder, beat with a mixer until smooth, then add pumpkin pureé and mix well. Leave to cool for later use.

基本麵糰搓揉及家庭式發酵方法
Basic Dough Kneading and Family Fermentation Method

做法

1. 將高筋麵粉放入鋼盆內，然後將奶粉、砂糖、細鹽、酵母、雞蛋、溫水及奶油（或酥油）一起加入，鹽不能與酵母放在同一位置，以免影響酵母的活躍能力。

2. 待酵母浸泡約5分鐘（即用酵母可以不用浸泡），水的表面出現泡泡狀，即代表酵母已開始活躍，產生二氧化碳，可以使麵糰膨脹。用手將麵糰快速拌勻，然後搓揉至起筋且麵糰表面光滑。將麵糰慢慢拉開時，出現薄膜且有彈性，即代表麵糰已產生筋性。

3. 初學者搓揉時間大約需要15分鐘。

Method

1. Put high gluten flour into a basin, then add milk powder, sugar, fine salt, yeast, eggs, warm water and butter (or shortening). Salt and yeast could not be put togehter, so as not to affect the active ability of yeast.

2. Soak yeast for about 5 minutes (no need to soak instant yeast). When there are bubbles on the surface of water, that means yeast begins to be active and emit carbon dioxide and it can help the rise of the dough. Mix the dough by hands quickly, then knead until gluten forms and the surface is smooth. When pulling the dough slowly, a membrane is formed and it is elastic, that means gluten is formed.

3. Beginners knead for about 15 minutes.

將麵糰收摺成光滑表面，並進行第一次發酵

1. 將麵糰壓扁且搓揉成長形，將麵糰頂部各角內摺起少許，然後向內摺向底部，並將麵糰從底部向前滾。將麵糰向右轉90度，再重覆動作4-5次，最後將麵糰略收及推成圓形。

2. 將麵糰放回鋼盆內，在麵糰表面噴一層水，並在鋼盆表面蓋上一層保鮮膜，表面再多噴一層水。然後放在已墊上濕毛巾烤盤上。

3. 烤箱用100℃預熱1-2分鐘後關掉。（用餘溫的熱度幫助麵糰發酵）

4. 將麵糰連同烤盤放入烤箱底層（爐火已關掉），進行第一次發酵35-45分鐘或發酵至原來麵糰2-3倍便可。（期間可以每15-20分鐘將鋼盆拿出來，然後將爐溫加熱1-2分鐘，再將鋼盆放回烤箱，這樣確保烤箱內的溫度平均，幫助麵糰在時間內完成發酵過程。）

5. 將麵糰從烤箱取出，要先測試麵糰是否已完成發酵，可用沾滿高筋麵粉的手指插入麵糰中央，若麵糰發酵適當，凹洞不會收縮，即表示第一次發酵完成。相反，若凹洞迅速收縮或回彈，即發酵未完成，要繼續發酵5-15分鐘或檢查程序有無出錯。

6. 發酵完成後，將麵糰取出，用手輕輕將麵糰按扁及排氣。將麵糰四邊收入再摺成長條形，然後將麵糰平均分割成所需等份，即時將麵糰搓圓或拍摺成長形，放在桌上用保鮮膜及濕毛巾蓋好，鬆弛麵糰15-20分鐘。

7. 麵糰鬆馳完成後，便可以進行包餡及最後發酵。

Fold the dough into a smooth surface for the first fermentation

1. Flatten the dough and knead into a rectangular shape. Fold the corners of the dough at the top a little inward, then fold inward towards the bottom. Then roll the dough from the bottom towards the front. Turn the dough for 90° to the right, and then repeat for 4-5 times. Finally, shape the dough slightly and push into a ball.

2. Put the dough back to the basin, spray water onto the surface of the dough. Cover the basin with a sheet of plastic wrap, spray water onto the surface again. Then put into a baking tray lined with a wet towel.

3. Preheat the oven at 100℃ for 1-2 minutes and turn off. (Use the sweltering heat to help dough ferments)

4. Put the dough with the baking tray into the lower deck of the oven. (the oven is turned off) First fermentation for 35-45 minutes, or ferments until 2-3 times of the original size. (During the fermentation, take out the basin for every 15-20 minutes, then turn on the oven for 1-2 minutes and then put back to the basin.This will ensure that the oven temperature is even, and the completion of the dough fermentation process within the specified time.)

5. Take out the dough from the oven and test if the fermentation of dough has completed, insert the centre of the dough with fingers stained with high gluten flour. If fermentation of dough is completed appropriately, the hole made by fingers will not shrink, it means that the first fermentation is completed. On the contrary, if the hole made by fingers is contracted or rebound rapidly, that means fermentation is not completed. Continue to ferment for 5-15 minutes or check if any error occurs during the procedures.

6. When fermentation is completed, take out the dough. Flatten the dough gently and release gas from the dough. Fold four sides of dough inwards and fold into rectangular shape. Then divide the dough into the required equal portions. Shape the dough into balls or rods. Put the dough onto the table, cover with plastic wrap and damp towel, let the dough rest for 15-20 minutes.

7. When resting of dough is completed, add the filling and proceed to final fermentation.

第一次發酵完成
First fermentation finished

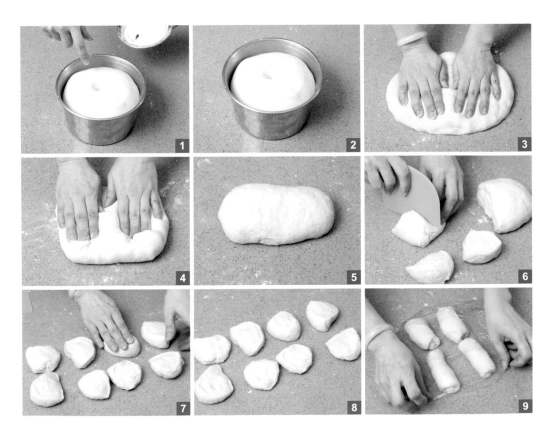

基本麵糰 - 甜麵糰（直接法）
Basic Dough - Sweet Dough (Direct Method)

麵糰材料 Dough Ingredients

高筋麵粉　250克/g
high gluten flour

無鹽奶油（或酥油）　15克/g
unsalted butter (or shortening)

奶粉　8克/g
milk powder

細鹽　2克/g
fine salt

溫水　135毫升/ml
warm water

砂糖　30克/g
sugar

雞蛋　25克/g
egg

即溶酵母　4 克/g
instant yeast

做法

1. 將高筋麵粉放入鋼盆內，然後將奶粉、砂糖、細鹽、酵母、雞蛋、溫水及奶油（或酥油）一起加入，鹽不能與酵母放在同一位置，以免影響酵母的活躍能力。

2. 待酵母浸泡約5分鐘（即溶酵母可以不用浸泡），水的表面出現泡泡狀，即代表酵母已開始活躍，產生二氧化碳，可以令麵糰膨脹。用手將麵糰快速拌勻，然後搓揉至起筋且麵糰表面光滑，將麵糰慢慢拉開時，出現薄膜及有彈性，即代表麵糰已產生筋性。初學者搓揉時間約15分鐘。

3. 將麵糰搓圓至表面光滑，放入鋼盆內，在麵糰表面噴一層水，蓋上保鮮膜，發酵40-45分鐘或發大至原來麵糰的2-3倍。

4. 將麵糰取出，要先測試麵糰是否已完成發酵，可用沾滿高筋麵粉的手指插入麵糰中央，若麵糰發酵適當，凹洞不會收縮，即表示第一次發酵完成。相反，若凹洞迅速收縮或回彈，即發酵未完成，要繼續發酵5-15分鐘或檢查程序有無出錯。

5. 發酵完成後，將麵糰取出，用手輕輕將麵糰按扁及排氣。將麵糰四邊收入再摺成長條形，然後將麵糰平均分割成所需等份，即時將麵糰搓圓或拍摺成長形，放在桌上用保鮮膜及濕布蓋好，鬆弛麵糰15-20分鐘。

6. 麵糰鬆弛後，將麵糰再次按扁排氣，然後包入餡料和塑形。將麵糰放在烤盤上，表面噴一層清水，蓋上保鮮膜進行最後發酵，發酵至麵糰發大2-3倍（約30-45分鐘）。

7. 發酵完成後，在麵包表面刷上全蛋液，放入已預熱180℃的烤箱烤15分鐘即可。

Method

1. Add high gluten flour into a basin, then add milk powder, sugar, fine salt, yeast, egg, warm water and butter (or shortening). Salt could not be put together with instant yeast so as not to affect the active capacity of instant yeast.

2. After yeast is soaked for about 5 minutes (instant yeast need not be soaked), there will be bubbles on the surface of the water. It means the yeast is already active and emitting carbon dioxide which could help the dough rising. Mix the dough by hands quickly, knead until gluten starts to form and surface of dough is smooth. When stretching the dough slowly, it is easy to pull out a small piece of transparent membrane and it is elastic. It means that gluten is generated. Beginners knead for about 15 minutes.

3. Knead the dough until the surface is smooth, put into a basin, spray some water on the surface of the dough, cover with plastic wrap and ferment for 45 minutes or until the dough rises in 2-3 times in volume.

4. Take out the dough and test if the fermentation is completed. Insert a finger stained with high gluten flour into the centre of the dough. If the dough is fermented appropriately, the hole will not shrink, it means that the first fermentation is completed. On the contrary, if the hole contracts or rebounds rapidly, that means fermentation is not completed, continue to ferment for 5-15 minutes or check if any errors occur during the procedures.

5. After fermentation is completed, take out the dough and press the dough to flatten to release gas gently. Fold the four sides inwards and fold into a strip. Divide the dough into equal portions, roll the dough in a ball or a rod instantly. Cover with plastic wrap and a damp cloth, let dough rest for 15-20 minutes.

6. After resting of dough, press the dough to flatten to release gas again. Wrap the filling and shape. Place the dough on a baking tray and spray some water on the surface, cover with plastic wrap for final fermentation until the dough rises in 2-3 times in volume (about 30-45 minutes).

7. After fermentation is completed, brush egg mixture on the surface of the bread. Bake in a preheated oven at 180°C for 15 minutes Serve.

基本麵糰 - 鹹麵糰（直接法）

Basic Dough - Sweet Dough (Direct Method)

麵糰材料 Dough Ingredients

高筋麵粉　250克/g
high gluten flour

無鹽奶油（或酥油）　20克/g
unsalted butter (or shortening)

奶粉　3克/g
milk powder

細鹽　5克/g
fine salt

溫水　165毫升/ml
warm water

砂糖　15克/g
sugar

即溶酵母　4 克/g
instant yeast

做法參考基本麵糰 - 甜麵糰（直接法）
Method refer to Basic Dough - Sweet Dough (Direct Method)

基本麵糰 – 湯種麵糰
Basic Dough - Utane Dough

湯種製法 （The method of making Utane starter）

高筋麵粉13克 +清水 65毫升（→ 製成品約有 67 克）13g high gluten flour + 65ml water → about 67g utane starter

1. 將高筋麵粉與清水拌勻，倒入小鍋內用慢火煮至濃稠。

2. 期間不停地用打蛋器或叉子攪拌至沒有粉粒，煮至糊狀及輕微滾起便可關火。

3. 將湯種倒入小碗中放涼一會，表面蓋上保鮮膜，防止麵糊變壞及水分流失。（建議在冰箱冷凍1-3小時後才用，效果會更佳。）

小提醒： 放在冰箱下層，保存期最好在5-6天內，或見表面變灰就要丟棄。

1. Mix well high gluten flour and water, pour into a small pot and cook over low heat and until thick.

2. Keep stirring with an egg whisk or fork and mix until smooth, cook until like a paste and slightly boil, turn off heat.

3. Put utane dough into a small bowl and leave to cool for a while, cover the surface with plastic wrap to prevent the batter from deterioration and loss of water. (Recommended to use after refrigerate for 1-3 hours, the result will be better.)

Note : Store in lower shelf of the refrigerator, preferably use within 5-6 days, or discard if the surface is dimmed.

主麵糰材料 Main Dough Ingredients

湯種1份 約67克/g 1 portion utane starter		溫水 90毫升/ml warm water	
高筋麵粉 250克/g high gluten flour		砂糖 35克/g sugar	
無鹽奶油 20克/g unsalted butter		雞蛋 25克/g egg	
奶粉 9克/g milk powder		即溶酵母 6克/g instant yeast	
細鹽 3克/g fine salt			

做法

1. 將高筋麵粉放入鋼盆內，然後將奶粉、砂糖、細鹽、酵母、雞蛋、湯種、溫水及無鹽奶油一起加入。

2. 待酵母浸泡約5分鐘，水的表面出現泡泡狀，即代表酵母已開始活躍。用手將麵糰快速拌勻，搓揉至起筋且麵糰表面光滑。初學者搓揉時間約15分鐘。

3. 將麵糰搓圓，放入鋼盆內，麵糰表面噴一層水，蓋上保鮮膜，發酵45分鐘或發大至原來麵糰的2-3倍。

4. 將麵糰取出，要預先測試麵糰是否已完成發酵，可用沾滿高筋麵粉的手指插入麵糰中央，若麵糰發酵適當，凹洞不會收縮，即表示第一次發酵完成。相反，若凹洞迅速收縮或回彈，即發酵未完成，要繼續發酵5-15分鐘或檢查程序有無出錯。

5. 發酵完成後，將麵糰取出，用手輕輕將麵糰按扁及排氣，將麵糰四邊收入再摺成長條形，然後將麵糰平均分割成所需等份，及即時將麵糰搓圓或拍摺成長形，放在桌上用保鮮膜及濕布蓋好，鬆弛麵糰15-20分鐘。

6. 麵糰鬆弛後，將麵糰再次按扁排氣，然後包入餡料及塑形。將麵糰放在烤盤上，表面噴一層清水，蓋上保鮮膜進行最後發酵，發酵至麵糰發大2-3倍（約30-45分鐘）。

7. 發酵完成後，在麵包表面刷上全蛋液，放入已預熱180℃的烤箱烤15分鐘即可。

Method

1. Add high gluten flour into a basin, then add milk powder, sugar, fine salt, yeast, egg, Utane starter, warm water and butter.

2. After yeast is soaked for about 5 minutes (instant yeast need not be soaked), there will be bubbles on the surface of the water. It means the yeast is already active. Mix the dough by hands quickly, knead until gluten starts to form and surface of dough is smooth. Beginners knead for about 15 minutes.

3. Knead the dough into a ball, put into a basin, spray some water on the surface of the dough, cover with plastic wrap and ferment for 45 minutes or until the dough rises in 2-3 times in volume.

4. Take out the dough and test if the fermentation is completed. Insert a finger stained with high gluten flour into the centre of the dough. If the dough is fermented appropriately, the hole will not shrink, it means that the first fermentation is completed. On the contrary, if the hole contracts or rebounds rapidly, that means fermentation is not completed, continue to ferment for 5-15 minutes or check if any errors occur during the procedures.

5. After fermentation is completed, take out the dough and press the dough to flatten to release gas gently. Fold the four sides inwards and fold into a strip. Divide the dough into equal portions, roll the dough in a ball or a rod instantly. Cover with plastic wrap and a damp cloth, let dough rest for 15-20 minutes.

6. After resting of dough, press the dough to flatten to release gas again. Wrap the filling and shape. Place the dough on a baking tray and spray some water on the surface, cover with plastic wrap for final fermentation until the dough rises in 2-3 times in volume (about 30-45 minutes).

7. After fermentation is completed, brush egg mixture on the surface of the bread. Bake in a preheated oven at 180° for 15 minutes . Serve.

小提醒 Tips

1. 湯種麵糰的特性較一般麵糰濕黏，搓時麵糰似泥狀不能成糰，因此需要用塑膠刮板輔助。
2. 搓完的麵糰表面較黏濕，可以灑少許高筋麵粉在桌上，讓麵糰表面沾上乾粉才容易揉圓。

1. Utane dough is wet and sticky compared with general dough. When kneading the dough, it is mud-like and not easy to form a dough, may need help with a scraper.

2. The surface of the kneaded dough is sticky and wet, sprinkle some high gluten flour onto the table. It is easy to knead if the surface of the dough is coated with dry powder.

低溫發酵法
Low Temperature Fermentation Method

所謂低溫發酵方法，是將剛搓揉好的麵糰馬上放入保鮮袋內綁緊，放入冰箱冷藏室（非結冰）12-24小時。如超過24小時（最多不超過72小時），就要將麵糰排出空氣，將保鮮袋綁緊，改用低溫的方法發酵。

待第二天做麵包時，將已發酵的麵糰取出按照當天室內氣溫，放於室溫下30-60分鐘。若當天的氣溫較熱時，約放30分鐘，天氣較寒冷時，可以放60分鐘。

然後按照正常的程序做麵包，即進行分割、鬆弛及塑形。

用這樣的方法發酵做出來的麵包，體積會較原來的小，但麵包內部組織較鬆軟，無需加入任何湯種，麵包的老化期也會延長。

但麵糰取出後，不同之處在於麵糰質地較為柔韌，塑形時要慢慢處理。

這個低溫發酵方法同樣地可以使用在已造型的麵包，將要進入最後發酵（已塑形）的整盆麵糰用保鮮膜覆蓋好，立即放入冰箱冷藏室（非結冰）8-10小時，待第二天要準備烤製前，預先放在室溫下 30-60分鐘（視乎天氣冷暖），再刷上全蛋液，然後送進烤箱。（但切記不可以將同一麵糰進行兩次低溫發酵）。

The so-called low temperature fermentation method is to put the kneaded dough into a plastic storage bag and fasten tightly immediately and then put into the refrigerator (non-freezing) for 12 to 24 hours. If more than 24 hours (minimum 72 hours), it is necessary to release gas from the dough, fasten the plastic storage bag tightly, then use low temperature fermentation method.

Until the next day to make the bread, take out the fermented dough and leave for 30-60 minutes according to room temperature. If the temperature is higher, leave for around 30 minutes, while the weather is colder, leave for around 60 minutes.

Then follow the procedures of making bread - separate, rest and style.

The volume of bread by using this fermentation method will be smaller, but the internal tissue of bread will be softer without adding any Utane starter, the aging period of bread will extended.

The texture of dough will be softer and it may need more time to shape and style.

The low temperature fermentation method can be used for the bread already shaped. Cover the whole tray of shaped bread with plastic wrap for final fermentation. Place in the refrigerator (non-freezing) immediately for 8-10 hours. Until before baking on the next day, leave at room temperature for 30-60 minutes (depends on whether the weather is warm or cold). Then brush with egg mixture and bake. (Remember not to use low temperature fermentation method twice for the same dough.)

最後發酵法

Final Fermentation Method

蒸籠發酵方法（竹製蒸籠）

　　一般家庭式烤箱，大部分都會選購18公升至23分升的家用烤箱。若要烤製6個或以上的麵包，可能要分2次烤製，而且需要有足夠地方進行發酵。

　　如果大家使用筆者提供的烤箱發酵方法，但家用烤箱太小時，未必能將整盆麵糰發酵。大家可能要將半份的麵糰改用蒸鍋發酵，但此舉動太麻煩了。

　　即使有更大的烤箱，筆者相信效果都不如蒸籠發酵的方法來得輕鬆及方便。

　　此方法可以同時處理 6-8 個麵糰，而且可以將蒸籠放在任何一個器皿或蒸鍋上，就可以利用水蒸氣上升的濕度及熱力，讓麵糰吸收並發酵至理想的效果。經常做麵包的朋友，只需要選購一組蒸籠，就像多了一個家庭式發酵箱一樣。不論任何天氣（包括其他國家天氣），只要懂得自行調整，就能在短時間內將麵糰發酵至理想的效果。

發酵方法

1. 將已塑形的麵糰分別墊上鋁箔紙，放入蒸籠內。（每個蒸籠最多放4個）

2. 8個麵糰就用2個蒸籠層疊在一起，蓋上竹製的蒸籠蓋。

3. 將1公升的滾水，倒在任何一個可以將蒸籠疊上，而且不會有空隙流出水蒸氣的器皿上即可。

4. 將兩個蒸籠疊放在熱水上，10分鐘後將兩個蒸籠上下調換位置，全程時間30分鐘。（即調換後還有20分鐘發酵時間）

5. 蒸籠最好放在廚房或溫暖且不通風的地方，這樣方便處理也不影響廚房工作。

6. 在發酵完成前，緊記預先將烤箱預熱，待麵糰完成最後發酵後，將麵糰取出，排放在烤盤上，刷上全蛋液便可以馬上送入烤箱，方便快捷。

7. 但要注意如果天氣較寒冷，室溫太低，中途（約15 分鐘）可能要將蒸籠暫時移開，將水煮滾後，再放回蒸籠繼續發酵。

8. 此另類的發酵方法，緊記切勿在麵糰上水，因為此方法提供大量較熱的水蒸氣，若麵包表面已噴水，麵糰會過濕而不能膨脹或成形。蒸籠中間也不需蓋上保鮮膜，否則下層水蒸氣會無法上升。由於是用竹製蒸籠，所以不用怕會有凝結的水氣回滴在麵包表面。若使用其他蒸籠，則要在最上層的麵糰蓋上保鮮膜防止凝結的水氣回滴。

好處

1. 在指定時間內，麵糰可發酵至理想。

2. 用具簡單，而蒸籠平日可用作蒸菜。

3. 只用熱開水1公升，使用任何一個輔助器皿盛裝熱水，便可以將蒸籠放在上方，進行最後發酵。

4. 發酵地方只要是溫暖的地方，室內任何地方都可以，不會阻礙廚房空間或阻礙煮飯。

5. 若無竹蒸也可以進行，只要略為注意水氣凝結的問題。

Steamer Fermentation Method (Bamboo Steamer)

Most of the family oven has 18-23 liter capacity. It may need to bake 6 breads in 2 turns. It also needs space for fermentation. If oven fermentation method is used, it may not be able to ferment the whole basin of bread when the oven is too small. You may use a steam wok to ferment half of the dough but this is troublesome.

Even if there is a bigger oven, it is believed that the result of steamer fermentation method is easier and more convenient.

This method can handle 6-8 doughs at the same time. The steamer can be placed in any container or steam wok. It allows the dough to absorb and have the best ferment result by utilizing humidity and heat by the rising steam. For those friends of bread, you could have a ferment box at home by buying a set of steamer. Regardless of any weather (including weather of other countries), as long as we know how to adjust, the desired results of fermentation could be obtained in a short period of time

Method of Fermentation

1. Arrange the shaped dough which lined with aluminum foil into the steamer. (Four doughs in each steamer the most.)

2. If there are 8 doughs, arrange in 2 steamers stacked together, cover with a bamboo lid.

3. Heat 1 liter of boiling water, pour into a container which 2 steamers could be stacked on and there is no gap for the steam to emit.

4. Put 2 steamers over the hot water, swap the upper and lower position of the steamers 10 minutes later, the entire period is 30 minutes. (i.e. 20 minutes of fermentation time after swaping)

5. It is best to place the steamers in a kitchen or a warm and unwindy place, which can facilitate the process and do not affect other tasks in the kitchen.

6. Before fermentation is completed, remember to preheat oven. After the final fermentation of the dough is completed. Take the dough out and arrange onto a baking tray. Brush with egg mixture and put into the oven instantly. It is convenient and time-saving.

7. If the room temperature is too low in cold weather, it may need to remove the steamer temporarily (about 15 minutes). When water is boiling, put the steamer back and continue to ferment.

8. Do not spray water onto the dough when using this alternative method of fermentation. Because this method provides a lot of hot steam, if the dough have been sprayed, it will be too wet and cannot rest or shape. The middle of the steamer need not be covered with plastic wrap, o r the steams in lower deck cannot be asceneded. Since the steamer is made by bamboo, the water will not sweat back to the dough. If other steamer is used, cover the dough on the top deck with plastic wrap to avoid water sweating back.

Merits

1. The dough will have ideal fermentation within the specified time.

2. Appliance is simple, the steamer can be used for steaming other dishes.

3. Only one liter of boiling water and any container for hot water are used, the steamer could be put for final fermentation.

4. Any warm place is suitable for fermentation, it will not obstruct the kitchen or affect the cooking of meals.

5. If there is no bamboo steamer, this method can still be used, the only concern is to adjust the sweating problem.

開始烘培
Start Baking

沙拉起士熱狗麵包
Salad And Cheese Sausage Bun

材料　Ingredients

高筋麵粉　250克/g
high gluten flour

無鹽奶油　15克/g
unsalted butter

奶粉　8克/g
milk powder

細鹽　2克/g
fine salt

溫水　135毫升/ml
warm water

砂糖　30克/g
sugar

雞蛋　25克/g
egg

即溶酵母　4克/g
instant yeast

餡料　Filling

熱狗　8條
sausages

馬自瑞拉起士　適量
some mozzarella

沙拉醬　適量
some salad dressing

表面　Topping

全蛋液（刷表面）　適量
some egg mixture for glazing

做法　Method

1. 詳細的基本麵包搓揉方法及第一次發酵方法請參閱 p.27。
2. 將麵糰平均分割成 8 份，拍摺成長形後，讓麵糰鬆弛 15-20 分鐘。
3. 將麵糰按壓成橢圓形，在麵糰中央按壓一道凹痕，將熱狗放在凹痕並按實。
4. 將麵包排放在烤盤上，進行最後發酵。（最後發酵方法請參閱 p.35）
5. 待麵糰發酵至兩倍大，麵包表面刷上全蛋液，擠上沙拉醬並灑上少許起士，放入已預熱的烤箱，用 180℃烤 15 分鐘即可。

1. For basic method of bread kneading and first fermentation, refer to p.27.
2. Divide the dough into 8 portions equally, roll into rods, let the dough rest for 15-20 minutes.
3. Press the dough and shape into oval shapes, press a pit in the centre of the dough, place a sausage onto the pit and press firmly.
4. Place the doughs onto a baking tray for final fermentation. (Refer to method of final fermentation on p.35)
5. Wait until the dough rises 2 times in volume, brush the surface with egg mixture, pipe salad dressing and sprinkle with shredded cheese. Bake in a preheated oven at 180°C for 15 minutes. Serve.

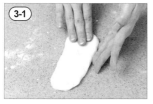

 小提醒 Tips

擺放熱狗時要平放並用力壓實，否則最後發酵後，熱狗歪掉了便不能再按壓。
Sausages should be placed flatly and pressed with force, otherwise they will be tilted after final fermentation and could no longer be pressed again.

芝麻熱狗麵包
Sesame And Sausage Bun

材料　Ingredients

高筋麵粉　250克/g
high gluten flour

無鹽奶油　15克/g
unsalted butter

奶粉　8克/g
milk powder

細鹽　2克/g
fine salt

溫水　135毫升/ml
warm water

砂糖　30克/g
sugar

雞蛋　25克/g
egg

即溶酵母　4克/g
instant yeast

餡料　Filling

熱狗　8 條
sausages

表面　Topping

全蛋液（刷表面）　適量
some egg mixture for glazing

做法　Method

1. 詳細的基本麵包搓揉方法及第一次發酵方法請參閱 p.27。
2. 將麵糰平均分割成 8 份，拍摺成長形後，讓麵糰鬆弛 15-20 分鐘。
3. 將麵糰按扁並按壓成一條繩子狀，然後將熱狗捲起，兩旁留 1 吋左右不用捲。
4. 將麵包排放在烤盤上，進行最後發酵。（最後發酵方法請參閱 p.35）
5. 待麵糰發酵至兩倍大，麵包表面刷上全蛋液並灑上少許芝麻，放入已預熱的烤箱，用 180℃烤 15 分鐘即可。

1. For basic method of bread kneading and first fermentation, refer to p.27.
2. Divide the dough into 8 portions equally, roll into rods, let the dough rest for 15-20 minutes.
3. Press the dough to flatten into a rope-like shape, roll up the sausage, leave 1 inch margin on both sides, no need to roll up.
4. Place the doughs onto a baking tray for final fermentation. (Refer to method of final fermentation on p.35)
5. Wait until the dough rises 2 times in volume, brush the surface with egg mixture and sprinkle with sesame seeds. Bake in a preheated oven at 180°C for 15 minutes. Serve.

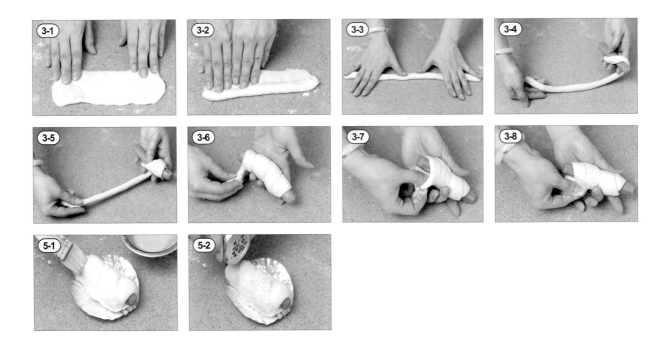

 小提醒 Tips

請勿將零碎的麵糰摺疊一起用來捲熱狗麵包，否則麵包可能爆裂。
Do not use a few bits of dough to roll up the sausage bun, or the bun will break up easily.

午餐肉麵包
Luncheon Meat Bun

材料　Ingredients

高筋麵粉　250克/g
high gluten flour

無鹽奶油　15克/g
unsalted butter

奶粉　8克/g
milk powder

細鹽　2克/g
fine salt

溫水　135毫升/ml
warm water

砂糖　30克/g
sugar

雞蛋　25克/g
egg

即溶酵母　4克/g
instant yeast

餡料　Filling

午餐肉　8 片/slices
luncheon meat

表面　Topping

全蛋液（刷表面）　適量
some egg mixture for glazing

白芝麻（表面裝飾）　適量
white sesame seeds for decoration

做法　Method

1. 詳細的基本麵包搓揉方法及第一次發酵方法請參閱 p.27 。
2. 將麵糰平均分割成 8 份，拍摺成長形後，讓麵糰鬆弛 15-20 分鐘。
3. 將麵糰按扁並輕輕拉長，將一片午餐肉放在麵糰的中央，麵糰上方摺向中央，下方向中央摺疊並按實，將麵包反轉，收口向下。
4. 將麵包放在烤盤上，進行最後發酵。（最後發酵方法請參閱 p.35）
5. 待麵糰發酵至兩倍大，麵包表面輕刷全蛋液並灑上芝麻作裝飾，放入已預熱的烤箱，用 180℃ 烤 15 分鐘即可。

1. For basic method of bread kneading and first fermentation, refer to p.27.
2. Divide the dough into 8 portions equally, roll into rods, let the dough rest for 15-20 minutes.
3. Press the dough to flatten and stretch slightly, place a slice of luncheon meat in the centre of the dough. Roll the upper part of the dough towards the centre, then roll the lower part of the dough towards the centre and press firmly. Invert the dough and place the openings at the bottom.
4. Place the doughs onto a baking tray for final fermentation. (Refer to method of final fermentation on p.35)
5. Wait until the dough rises 2 times in volume, brush the surface with egg mixture and sprinkle with sesame seeds for decoration. Bake in a preheated oven at 180°C for 15 minutes. Serve.

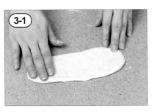

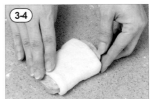

 小提醒 Tips

收口一定要用力壓實，否則，發酵後收口容易裂開。
The openings should be pressed firmly, otherwise, they will be broken easily after fermentation.

白醬雞肉麵包
Bun With Chicken In White Sauce

材料　Ingredients

高筋麵粉　250克/g
high gluten flour

奶粉　8克/g
milk powder

酥油　15毫升/ml
shortening

細鹽　2克/g
fine salt

溫水　135毫升/ml
warm water

砂糖　30克/g
sugar

雞蛋　25克/g
egg

即溶酵母　4克/g
instant yeast

餡料　Filling

白醬雞肉餡（做法請參閱p.18）
Chicken in white sauce filling
(Refer to the method on p.18)

表面　Topping

全蛋液（刷表面）　適量
some egg mixture for glazing

做法　Method

1. 詳細的基本麵包搓揉方法及第一次發酵方法請參閱 p.27。
2. 將麵糰平均分割成 8 份，滾圓後，讓麵糰鬆弛 15-20 分鐘。
3. 將麵糰按扁，包入適量雞肉餡，或在收口前改作其他造型，黏緊收口，將麵糰翻轉收口向下，用手輕輕按壓麵糰。
4. 將麵包排放在烤盤上，進行最後發酵。（最後發酵方法請參閱 p.35）
5. 待麵糰發酵至兩倍大，在表面輕刷全蛋液，並灑上少許香草裝飾，放入已預熱的烤箱，用 180℃烤 15 分鐘即可。

1. For basic method of bread kneading and first fermentation, refer to p.27.
2. Divide the dough into 8 portions equally, roll into balls, let the dough rest for 15-20 minutes.
3. Press the dough to flatten, put in right amount of filling. Or shape into other shapes. Press the openings firmly. Invert the dough and place the openings at the bottom. Press the dough slightly.
4. Place the doughs onto a baking tray for final fermentation. (Refer to method of final fermentation on p.35)
5. Wait until the dough rises 2 times in volume, brush the surface with egg mixture and sprinkle with some herbs for decoration. Bake in a preheated oven at 180°C for 15 minutes. Serve.

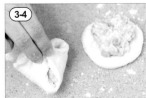
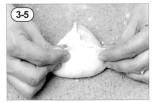

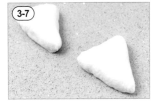

 小提醒 Tips

　　白醬雞肉麵包可以做出不同的形狀，只要在收口前構想要做的形狀，便能得心應手。
Bun with Chicken in White Sauce could be in different shapes, as long as shaping the dough before pressing the openings.

叉燒餐包

Bun With Barbecued Pork

材料　Ingredients

高筋麵粉　250克/g
high gluten flour

奶粉　8克/g
milk powder

砂糖　30克/g
sugar

細鹽　2克/g
fine salt

無鹽奶油（或酥油）
15克/g
unsalted butter

溫水　35毫升/ml
warm water

雞蛋　25克/g
egg

即溶酵母　4克/g
instant yeast

餡料　Filling

叉燒餡（請參照 p.19 的製法）
Barbecued pork filling
(Refer to the method on p.19)

表面糖漿
Syrup for Brushing the Surface

熱水　1湯匙/tbsp
hot water

砂糖（預先拌溶）　2茶匙/tsps
sugar (stir until dissolve in advance)

做法　Method

1. 詳細的基本麵包搓揉方法及第一次發酵方法，請參閱 p.27 。
2. 將麵糰平均分割成 8 份，滾圓後，讓麵糰鬆弛 15-20 分鐘。
3. 將麵糰按扁，包入適量叉燒餡，黏緊收口，將麵糰翻轉收口向下，用手輕輕按壓麵糰。
4. 將麵包放在烤盤上，進行最後發酵。（參考最後發酵方法 p.35）
5. 待麵糰發酵至兩倍大，在表面輕刷全蛋液，放入巳預熱的烤箱，用 180℃ 烤 13 分鐘，
   出爐後刷上一層糖漿，放回入烤箱再烤 2 分鐘即可。

1. For basic method of bread kneading and first fermentation, refer to p.27.
2. Divide the dough into 8 portions equally, roll into balls, let the dough rest for 15-20 minutes.
3. Press the dough to flatten, put in right amount of barbecued pork filling. Press the openings firmly. Invert the dough and place the openings at the bottom. Press the dough slightly.
4. Place the doughs onto a baking tray for final fermentation. (Refer to method of final fermentation on p.35)
5. Wait until the dough rises 2 times in volume, brush the surface with egg mixture slightly. Bake in a preheated oven at 180°C for 13 minutes. Take out. Brush the surface with syrup. Bake for 2 more minutes. Serve.

 小提醒 Tips

出爐後的麵包除了刷糖漿外，還可以改刷蜜糖，但蜜糖要用少許熱水稀釋。
Instead of brushing syrup on the surface, honey could be used, however, honey need to be diluted with some hot water before brushing.

香草玉米牛肉麵包
Corn And Beef Bun With Herbs

材料　Ingredients

高筋麵粉　250克/g
high gluten flour

酥油　15克/g
shortening

奶粉　8克/g
milk powder

細鹽　2克/g
fine salt

溫水　135毫升/ml
warm water

砂糖　30克/g
sugar

雞蛋　25克/g
egg

即溶酵母　4克/g
instant yeast

餡料　Filling

香草玉米牛肉餡
（請參照 p.20 的製法）
Corn and beef filling with herbs
(Refer to the method on p.20)

表面　Topping

全蛋液（刷表面）　適量
some egg mixture for glazing

香草（表面裝飾）　適量
some herbs for decoration

做法　Method

1. 詳細的基本麵包搓揉方法及第一次發酵方法請參閱 p.27。
2. 將麵糰平均分割成 8 份，滾圓後，讓麵糰鬆弛 15-20 分鐘。
3. 將麵糰按扁，包入適量香草玉米牛肉餡，黏緊收口，將麵糰反轉收口向下，用手輕輕按壓麵糰。
4. 將麵包排放在烤盤上，進行最後發酵。（最後發酵方法請參閱 p.35）
5. 待麵糰發酵至兩倍大，在表面輕刷全蛋液，並灑上少許香草裝飾，放入已預熱的烤箱，用 180℃ 烤 15 分鐘即可。

1. For basic method of bread kneading and first fermentation, refer to p.27.
2. Divide the dough into 8 portions equally, roll into balls, let the dough rest for 15-20 minutes.
3. Press the dough to flatten, put in right amount of corn and beef filling with berbs. Press the openings firmly. Invert the dough and place the openings at the bottom. Press the dough slightly.
4. Place the doughs onto a baking tray for final fermentation. (Refer to method of final fermentation on p.35)
5. Wait until the dough rises 2 times in volume, brush the surface with egg mixture slightly. Sprinkle with some herbs for decoration. Bake in a preheated oven at 180℃ for 15 minutes. Serve.

小提醒 Tips

餡料一定要瀝乾水分，收口時一定要小心，切勿放太多餡料。
Filling must be drained thoroughly, pay attention when pressing the openings and do not put in too much filling.

皇冠杏仁麵包
Crown Almond Bun

用具　Utensils

花紋戚風蛋糕模（6吋）
Chiffon cake mold with
braided pattern (6")

材料　Ingredients

高筋麵粉　250克/g
high gluten flour

無鹽奶油　15克/g
unsalted butter

奶粉　8克/g
milk powder

細鹽　2克/g
fine salt

溫水　135毫升/ml
warm water

砂糖　30克/g
sugar

雞蛋　25克/g
egg

即溶酵母　4克/g
instant yeast

餡料　Filling

杏仁粒（已烤脆）　30克/g
almonds (baked till crispy)

表面　Topping

全蛋液（刷表面）　適量
some egg mixture for glazing

杏仁粒及杏仁片
（表面裝飾）　適量
chopped almonds and almond
flakes for decoration

做法　Method

1. 詳細的基本麵包搓揉方法及第一次發酵方法請參閱 p.27 。
2. 將杏仁粒加入麵糰內揉勻。
3. 先將麵糰平均分為 2 份。再平均分為 5 份，每份約 45 克，拍摺成長形，讓麵糰鬆弛 15-20 分鐘。
4. 將麵糰按扁並輕輕拉長，將下方 1/3 部分麵糰摺向中央，麵糰上方摺向中央，再捲向底部，將收口壓實，搓揉成橄欖形。
5. 模具預先刷油，將麵包平均排放入模具內，進行最後發酵。
6. 待麵糰發酵至兩倍大，在表面輕刷上全蛋液，再放上杏仁片，放入已預熱的烤箱，用 180℃烤 20 分鐘即可。

1. For basic method of bread kneading and first fermentation, refer to p.27.
2. Add chopped almonds into the dough and knead well.
3. Divide the dough into 2 portions, then divide each portions into 5 small portions equally, each weighs 45g approximately. Roll into rods, let the dough rest for 15-20 minutes.
4. Press the dough to flatten and stretch slightly, fold the lower 1/3 part of dough towards the centre, folder the upper part towards the centre. Then roll towards the bottom. Press the openings firmly. Shape into an olive.
5. Brush the mold with oil, arrange the doughs onto the mold evenly for final fermentation.
6. Wait until the dough rises 2 times in volume, brush the surface with egg mixture slightly. Sprinkle with almond flakes. Bake in a preheated oven at 180°C for 20 minutes. Serve.

小提醒 Tips

1. 剩下另一半麵糰，可分成2份，每份切成3條，一端不要完全切斷。
2. 將麵糰左右交叉編織成辮子狀，收口壓實。放上杏仁片即為杏仁條。

1. Divide the remaining half of dough into 2 portions. Cut each into 3 strips but do not cut off.
2. Cross weave into braid-like shape, press the opening. Top with almond flakes to make Almond Stick.

培根起士條
Bacon And Cheese Stick

材料　Ingredients

高筋麵粉　250克/g
high gluten flour

奶粉　8克/g
milk powder

細鹽　2克/g
fine salt

酥油　15克/g
shortening

溫水　135毫升/ml
warm water

雞蛋　25克/g
egg

即溶酵母　4克/g
instant yeast

餡料　Filling

培根　適量
some bacon

馬自瑞拉起士　適量
some mozzarella

表面　Topping

全蛋液（刷表面）　適量
some egg mixture for glazing

做法　Method

1. 詳細的基本麵包搓揉方法及第一次發酵方法，請參閱 p.27。

2. 將麵糰平均分割成 8 份，拍摺成長形後，讓麵糰鬆弛 15-20 分鐘。

3. 將麵糰按扁並輕輕拉長，平均分割成 3 條長條形，頂部不用切段。將麵糰從左邊疊放在中央麵糰，然後右邊又疊向左邊（中央）麵糰，重覆這動作，左疊右，右疊左。收口壓實。將培根餡料放在麵包表面。

4. 將麵包排放在烤盤上，進行最後發酵。（參考最後發酵方法 p.35）

5. 待麵糰發酵至兩倍大，麵包表面刷上全蛋液，然後灑上起士碎粒，放入巳預熱的烤箱，用 180℃烤 15 分鐘至起士呈現金黃色即可。

1. For basic method of bread kneading and first fermentation, refer to p.27.

2. Divide the dough into 8 portions equally, roll into rods, let the dough rest for 15-20 minutes.

3. Flatten the dough and stretch gently, divide into 3 long strips evenly but do not cut off. Stack the dough from the left towards the centre, then stack the dough from right towards the centre, repeat this action. Press firmly. Top with bacon filling.

4. Place the doughs onto a baking tray for final fermentation. (Refer to method of final fermentation on p.35)

5. Wait until the dough rises 2 times in volume, brush the surface with egg mixture, sprinkle with shredded cheese. Bake in a preheated oven at 180°C for 15 minutes until cheese is golden brown. Serve.

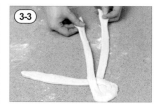

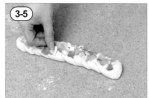

小提醒 Tips

1. 起士可以改用帕瑪森起士或任何起士都可以。
2. 表面可以改放杏仁片做成杏仁條。
1. Parmesan or other cheese can be used instead.
2. It could be topped with almond flakes to make Almond Stick.

53

榛子巧克力扭紋條
Hazelnut And Chocolate Roll With Twisted Pattern

材料　Ingredients

高筋麵粉　250克/g
high gluten flour

酥油　15克/g
shortening

奶粉　8克/g
milk powder

細鹽　2克/g
fine salt

溫水　135毫升/ml
warm water

砂糖　30克/g
sugar

雞蛋　25克/g
egg

即溶酵母　4克/g
instant yeast

餡料　Filling

榛子巧克力醬　2-3湯匙
hazelnut and chocolate
sauce

杏仁粒（表面）　適量
some chopped almonds
for toppings

表面　Topping

全蛋液（刷表面）　適量
some egg mixture for glazing

做法　Method

1. 詳細的基本麵包搓揉方法及第一次發酵方法請參閱 p.27。

2. 將麵糰平均分割成 2 份，拍摺成長形後，讓麵糰鬆弛 15-20 分鐘。

3. 將麵糰用木棍桿薄成 7 吋 ×12 吋長形，將榛子巧克力醬平均塗在麵糰上，將麵糰對摺並將四邊壓實。壓成約 7 吋 ×7 吋方形麵糰，然後將麵糰平均切成 4-5 份長條形。在每條麵糰中央切一刀，將麵糰向左右兩邊扭成紐紋狀並打結成花捲。

4. 將花捲排放在烤盤上，進行最後發酵。（最後發酵方法請參閱 p.35）

5. 待麵糰發酵至兩倍大，在表面輕輕刷上全蛋液，放入已預熱的烤箱，用 180℃烤 15 分鐘至表面呈現金黃色即可。

1. For basic method of bread kneading and first fermentation, refer to p.27.

2. Divide the dough into 2 portions equally, roll into rods, let the dough rest for 15-20 minutes.

3. Roll the dough with a rolling pin into a 7 inch x 12 inch rectangle. Spread hazelnut and chocolate sauce onto the dough evenly. Fold the dough in halves and press the 4 sides firmly. The dough should be about 7 inch x 7 inch square. Then cut the dough into 4-5 strips evenly. Slit in the centre of each strip, then twist the strips towards opposite directions into twisted shape and tie a knot shaped like a flower.

4. Place the flower-shaped doughs onto a baking tray for final fermentation. (Refer to method of final fermentation on p.35)

5. Wait until the dough rises 2 times in volume, brush the surface with egg mixture. Bake in a preheated oven at 180°C for 15 minutes. Serve.

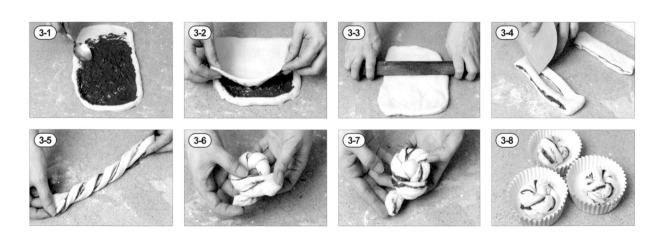

小提醒 Tips

餡料可以改用30克巧克力，放在熱水隔水加熱，再加上杏仁或榛子粒代替榛子醬。
Hazelnut and chocolate sauce could be replaced by melting 30g of chocolate block with chopped almonds or hazelnuts.

椰香捲
Coconut Bun

材料　Ingredients

高筋麵粉　250克/g
high gluten flour

無鹽奶油 (或酥油)
15克/g
unsalted butter
(or shortening)

奶粉　8克/g
milk powder

細鹽　2克/g
fine salt

溫水　135毫升/ml
warm water

砂糖　30克/g
sugar

雞蛋　25克/g
egg

即溶酵母　4克/g
instant yeast

餡料　Filling

椰香餡（請參照 p.20 的製法）
Shredded coconut filling
(Refer to the method on p.20)

表面　Topping

全蛋液（刷表面）　適量
some egg mixture for glazing

做法　Method

1. 詳細的基本麵包搓揉方法及第一次發酵方法請參閱 p.27。

2. 將麵糰平均分割成 2 份，拍摺成長形後，讓麵糰鬆弛 15-20 分鐘。

3. 將麵糰用木棍桿薄成 6 吋 ×12 吋長形，將椰香餡平均放在麵糰上，麵糰由上向下捲，收口略為拉薄並壓實。然後分割成所需厚度或等份，或按照要做的模具大小，預先在分割麵糰時決定要做的份量。

4. 將麵包放在烤模內，然後放在烤盤上，進行最後發酵。（最後發酵方法請參閱 p.35）

5. 待麵糰發酵至兩倍大，輕手掃上全蛋液，放入已預熱的烤箱，用 180℃ 烤 15 分鐘至表面呈現金黃色即可。

1. For basic method of bread kneading and first fermentation, refer to p.27.

2. Divide the dough into 2 portions equally, roll into rods, let the dough rest for 15-20 minutes.

3. Roll the dough with a rolling pin into a 6 inch x 12 inch rectangle. Spread shredded coconut filling onto the dough evenly. Roll the dough from the top to the bottom. Make the tip thinner and press firmly. Then divide the dough into portions in same thickness evenly. Or decide the portions of dough according to the size of the mold.

4. Place the dough into the mold and put onto a baking tray for final fermentation. Refer to method of final fermentation on p.35)

5. Wait until the dough rises 2 times in volume, brush the surface with egg mixture. Bake in a preheated oven at 180°C for 15 minutes until golden brown. Serve.

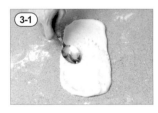
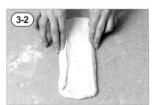
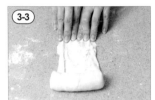

小提醒 Tips

椰香捲變化多，可以將一半麵糰桿成 9吋×5吋長型，捲起放入半磅方形面麵包模內。
或可將一半麵糰分成2份，桿成 3吋×6吋長型，捲起放入迷你長形模。

The shapes of Coconut Bun varies. Roll half of the dough into 9 inch x 5 inch rectangle, roll up and place into a half pound square bread mold. Or divide half of the dough into 2 portions and roll into a 3 inch x 6 inch rectangle, roll up and place into a mini long mold.

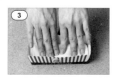

鮪魚蛋白麵包
Tuna And Egg White Bun

材料　Ingredients

高筋麵粉　250克/g
high gluten flour

酥油　15克/g
shortening

奶粉　8克/g
milk powder

細鹽　2克/g
fine salt

溫水　135毫升/ml
warm water

砂糖　30克/g
sugar

雞蛋　25克/g
egg

即溶酵母　4克/g
instant yeast

餡料　Filling

鮪魚魚蛋白餡
（請參照 p.23 餡料製法）
Tuna and Egg White Filling
(Refer to the method on p.23)

表面　Topping

全蛋液（刷表面）　適量
some egg mixture for glazing

馬自瑞拉起士碎粒　適量
some shredded mozzarella

做法　Method

1. 詳細的基本麵包搓揉方法及第一次發酵方法請參閱 p.27 。
2. 將麵糰平均分割成 8 份，滾圓，讓麵糰鬆弛 15-20 分鐘。
3. 將麵糰切出 1/3 麵糰，然後再分 2 份麵糰，搓成長條形，剩下的麵糰搓圓及按扁，將 2 條麵條分兩層，圍在圓形麵糰上，收口捏實，將餡料放在中央。
4. 將麵包排放在烤盤上，進行最後發酵。（最後發酵方法請參閱 p.35）
5. 待麵糰發酵至兩倍大，輕刷全蛋液，表面灑上少許起士碎粒，放入已預熱的烤箱，用 180℃烤 15 分鐘即可。

1. For basic method of bread kneading and first fermentation, refer to p.27.
2. Divide the dough into 8 portions equally, roll into balls, let the dough rest for 15-20 minutes.
3. Take out 1/3 of the dough and divide into 2 equal portions, roll into strips. Press the remaining dough to flatten into round shape. Arrange the strips around the side of the round dough one by one. Press the openings firmly. Place the filling in the centre.
4. Place the doughs onto a baking tray for final fermentation. (Refer to method of final fermentation on p.35)
5. Wait until the dough rises 2 times in volume, brush the surface with egg mixture, sprinkle with shredded cheese. Bake in a preheated oven at 180°C for 15 minutes. Serve.

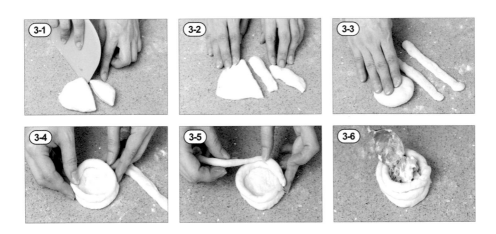

麵包造型可隨個人喜好，餡料可以改為包入麵糰內，做成任何形狀。
The shape of bun varies according to personal preferences, the filling can be placed inside the bun and made into any shape.

59

紫心番薯麵包
Purple Sweet Potato Bun

材料　Ingredients

高筋麵粉　250克/g
high gluten flour

酥油　15克/g
shortening

奶粉　8克/g
milk powder

細鹽　2克/g
fine salt

溫水　120毫升/ml
warm water

砂糖　30克/g
sugar

雞蛋　20克/g
egg

即溶酵母　4克/g
instant yeast

番薯泥　120克/g
sweet potato puree

餡料　Filling

紫心番薯餡　160克/g
（請參照 p.21 餡料製法）
sweet potato puree
(Refer to the method on p.21)

表面　Topping

糖粉、可可粉　各適量
some icing sugar and
cocoa powder

全蛋液(刷表面)　適量
some egg mixture for glazing

做法　Method

1. 詳細的基本麵包搓揉方法及第一次發酵方法請參閱 p.27。
2. 將麵糰平均分割成 8 份，滾圓後，讓麵糰鬆弛 15-20 分鐘。
3. 將麵糰按扁，包入適量番薯餡，黏緊收口及塑型，收口向下，用手輕輕按壓麵糰。
4. 將麵包排放在烤盤上，進行最後發酵。（最後發酵方法請參閱 p.35）
5. 待麵糰發酵至兩倍大，在表面輕刷全蛋液，放入已預熱的烤箱，用 180℃烤 15 分鐘即可。放涼後灑上糖粉及可可粉裝飾。

1. For basic method of bread kneading and first fermentation, refer to p.27.
2. Divide the dough into 8 portions equally, roll into balls, let the dough rest for 15-20 minutes.
3. Press the dough to flatten, put in right amount of sweet potato filling. Press the openings firmly. Invert the dough and place the openings at the bottom.
4. Place the doughs onto a baking tray for final fermentation. (Refer to method of final fermentation on p.35)
5. Wait until the dough rises 2 times in volume, brush the surface with egg mixture. Bake in a preheated oven at 180°C for 15 minutes. Leave to cool and sprinkle with icing sugar and cocoa powder for decoration. Serve.

 小提醒 Tips

番薯含高纖維，有助腸道的蠕動。
Sweet potatoes are with high amount of fibres which helps the wriggling of intestines.

火腿葱麵包
Ham And Onion Bun

材料　Ingredients

高筋麵粉　250克/g
high gluten flour

無鹽奶油　20克/g
unsalted butter

細鹽　5克/g
fine salt

奶粉　3克/g
milk powder

溫水　160毫升/ml
warm water

砂糖　5克/g
sugar

即溶酵母　4克/g
instant yeast

餡料　Filling

火腿葱餡料 （請參閱 p.23 的製法）
Ham and onion filling
（Refer to the method on p.23）

全蛋液（刷表面）　適量
some egg mixture for glazing

表面　Topping

全蛋液(刷表面)　適量
some egg mixture for glazing

做法　Method

1. 詳細的基本麵包搓揉方法及第一次發酵方法請參閱 p.27 。

2. 將麵糰平均分割成 4 份，拍摺成長形後，讓麵糰鬆弛 15-20 分鐘。

3. 將麵糰按扁並用木棍桿薄成長方形，將餡料平均鋪在麵糰上，然後將麵糰由上向下捲，捲至尾部時將麵糰拉薄，繼續捲並在收口處略微拉薄並壓實。將麵糰分割成兩份。

4. 將麵包排放在烤盤上，進行最後發酵。（最後發酵方法請參閱 p.35）

5. 待麵糰發酵至兩倍大，在表面輕刷上全蛋液，放入已預熱的烤箱，用 180℃烤 15 分鐘即可。

1. For basic method of bread kneading and first fermentation, refer to p.27.

2. Divide the dough into 4 portions equally, roll into rods, let the dough rest for 15-20 minutes.

3. Press the dough to flatten and roll into rectangular shape by a rolling pin, arrange filling onto the dough evenly. Roll the dough from the top to the bottom. Make the tip thinner and press firmly. Divide the dough into 2 equal portions.

4. Place the doughs onto a baking tray for final fermentation. (Refer to method of final fermentation on p.35)

5. Wait until the dough rises 2 times in volume, brush the surface with egg mixture. Bake in a preheated oven at 180°C for 15 minutes. Serve.

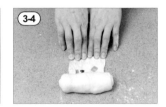

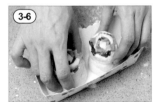

小提醒 Tips

麵糰造型可隨個人喜好或創意。麵糰亦可切成3份，收口不用切斷，交叉反轉摺疊成品字形狀。

The shape of the dough varies according to personal preferences or creativity. The dough can also be cut into 3 portions and do not cut off, then cross and turn into the shape of a Chinese word (like 3 circles).

香蒜麵包
Garlic Bread

材料　Ingredients

高筋麵粉　250克/g
high gluten flour

無鹽奶油　20克/g
unsalted butter

細鹽　5克/g
fine salt

奶粉　3克/g
milk powder

溫水　160毫升/ml
warm water

砂糖　15克/g
sugar

即溶酵母　4克/g
instant yeast

餡料　Filling

香蒜包餡料
（蒜蓉醬製法請參閱p.24）
The Pesto Filling
Refer to p.24 for the method

表面　Topping

全蛋液(刷表面)　適量
some egg mixture for glazing

做法　Method

1. 詳細的基本麵包搓揉方法及第一次發酵方法請參閱 p.27。
2. 將麵糰平均分割成 8 份，拍摺成長形後，讓麵糰鬆弛 15-20 分鐘。
3. 將麵糰按扁，由上向下捲，捲至尾部時將麵糰拉薄，繼續捲並將收口向下壓實。
4. 將麵包放在烤盤上，進行最後發酵。（參考最後發酵方法 p.35）
5. 待麵糰發酵至兩倍大，在表面輕刷上全蛋液，用刀片在麵糰中央快速劃一下後，抹上蒜蓉醬，放入已預熱的烤箱，用 180℃烤 15 分鐘即可。

1. For basic method of bread kneading and first fermentation, refer to p.27.
2. Divide the dough into 8 portions equally, roll into rods, let the dough rest for 15-20 minutes.
3. Press the dough to flatten. Roll the dough from the top to the bottom. Make the tip thinner and press firmly. Continue to roll, invert the dough and place the openings at the bottom.
4. Place the doughs onto a baking tray for final fermentation. (Refer to p.35 for method of final fermentation)
5. Wait until the dough rises 2 times in volume, brush the surface with egg mixture slightly. Slit in the centre of the dough by a blade quickly Spread pesto filling onto the dough.Bake in a preheated oven at 180°C for 15 minutes. Serve.

 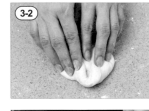 

 小提醒 Tips

若烤箱未預熱，切勿預早將麵糰塗全蛋液或切開表面。
If the oven has not preheated, do not brush egg mixture or slice the dough too early.

芝麻捲
Sesame Roll

材料　Ingredients

高筋麵粉　250克/g
high gluten flour

酥油　20克/g
shortening

細鹽　5克/g
fine salt

奶粉　3克/g
milk powder

溫水　160毫升/ml
warm water

砂糖　15克/g
sugar

即溶酵母　4克/g
instant yeast

表面　Topping

表面芝麻　適量
some sesame seeds

全蛋液（刷表面）　適量
some egg mixture
for glazing

做法　Method

1. 詳細的基本麵包搓揉方法及第一次發酵方法請參閱 p.27。
2. 將麵糰平均分割成 8 份，拍摺成長形後，讓麵糰鬆弛 15-20 分鐘。
3. 將麵糰按扁並拉成 T 字形，由上向下捲，捲至尾部時將麵糰拉薄，繼續捲並將收口向下壓實。
4. 將麵包放在烤盤上，進行最後發酵。（最後發酵方法請參閱 p.35）
5. 待麵糰發酵至兩倍大，在表面輕刷上全蛋液及灑上芝麻，放入已預熱的烤箱，用 180℃ 烤 15 分鐘即可。

1. For basic method of bread kneading and first fermentation, refer to p.27.
2. Divide the dough into 8 portions equally, roll into rods, let the dough rest for 15-20 minutes.
3. Press the dough to flatten and stretch to "T" shape. Roll the dough from the top to the bottom. Make the tip thinner and press firmly. Continue to roll, invert the dough and place the openings at the bottom.
4. Place the doughs onto a baking tray for final fermentation. (Refer to p.35 for method of final fermentation)
5. Wait until the dough rises 2 times in volume, brush the surface with egg mixture slightly. Sprinkle with sesame seeds. Bake in a preheated oven at 180°C for 15 minutes. Serve.

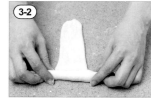

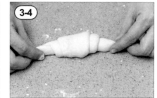

 小提醒 Tips

麵糰拉成 T 字形時，可以在表面塗上少許奶油才捲，吃時感覺更美味。但切忌塗太多，以免烘烤後收口裂開。

When stretching the dough into "T" shape, spread some butter on the surface before rolling would make it more delicious. But do not spread too much or the opening will be broken easily.

起士麵包
Cheese Bun

材料　Ingredients

高筋麵粉　250克/g
high gluten flour

酥油　20克/g
shortening

細鹽　5克/g
fine salt

奶粉　3克/g
milk powder

溫水　160毫升/ml
warm water

砂糖　15克/g
sugar

即溶酵母　4克/g
instant yeast

餡料　Filling

帕瑪森起士　適量
some parmesan cheese

沙拉醬　適量
some salad dressing

表面　Topping

全蛋液（刷表面）　適量
some egg mixture
for glazing

做法　Method

1. 詳細的基本麵包搓揉方法及第一次發酵方法請參閱 p.27 。
2. 將麵糰平均分割成 8 份，拍摺成長形後，讓麵糰鬆弛 15-20 分鐘。
3. 將麵糰按扁，由上向下捲，捲至尾部時將麵糰拉薄，繼續捲並將收口向下壓實。
4. 將麵包排放在烤盤上，進行最後發酵。（最後發酵方法請參閱 p.35）
5. 待麵糰發酵至兩倍大，在表面刷上全蛋液，擠上少許沙拉及灑上起士碎粒，放入已預熱的烤箱，用 180℃烤 15 分鐘即可。

1. For basic method of bread kneading and first fermentation, refer to p.27.
2. Divide the dough into 8 portions equally, roll into rods, let the dough rest for 15-20 minutes.
3. Press the dough to flatten. Roll the dough from the top to the bottom. Make the tip thinner and press firmly. Continue to roll and invert the dough and place the openings at the bottom.
4. Place the doughs onto a baking tray for final fermentation. (Refer to p.35 for method of final fermentation)
5. Wait until the dough rises 2 times in volume, brush the surface with egg mixture slightly. Pipe some salad dressing and sprinkle with chopped cheese. Bake in a preheated oven at 180°C for 15 minutes. Serve.

 小提醒 Tips

用帕瑪森起士，烤出來表面有金黃色的效果。改用馬自瑞拉起士也可以。
Parmesan cheese is used to make the golden effect on surface. Or replaced by mozzarella.

豬肉蛋滿福堡
Pork Loin And Egg Muffin

材料　Ingredients

高筋麵粉　200克/g
high gluten flour

奶粉　10克/g
milk powder

酥油　8克/g
shortening

細鹽　3克/g
salt

溫水　135毫升/ml
warm water

即溶酵母　4克/g
instant yeast

砂糖　4克/g
sugar

餡料　Filling

豬肉漢堡排　4塊/slices
pork loin

已蒸熟的雞蛋
（即太陽蛋）4個
steamed egg (i.e., sunny eggs)

表面　Topping

玉米粉　適量
some cornflour

做法　Method

1. 詳細的基本麵包搓揉方法及第一次發酵方法請參閱 p.27。
2. 將麵糰平均分割成 6 份，搓圓後，讓麵糰鬆弛 15-20 分鐘。
3. 麵糰再按壓排氣並搓圓，然後在麵包表面噴水，底部沾上玉米粉，輕手按扁。
4. 將麵包排放在烤盤上，進行最後發酵。（最後發酵方法請參閱 p.35）
5. 發酵完成後，在麵包表面蓋上一張鋁箔紙，另用一個長形或方形的烤盤輕壓在鋁箔紙上，放入已預熱 180℃的烤箱，用相同烤箱溫度烤約 15 分鐘，加入餡料即可。

1. For basic method of bread kneading and first fermentation, refer to p.27.

2. Divide the dough into 6 portions equally, roll into balls, let the dough rest for 15-20 minutes.

3. Press the dough again and roll into balls. Coat with cornflour. Flatten slightly.

4. Place the doughs onto a baking tray for final fermentation. (Refer to p.35 for method of final fermentation)

5. After final fermentation, cover the dough with a piece of aluminium foil. Top with a rectangular or square baking tray. Bake in a preheated oven at 180°C for 15 minutes. Sandwich with filling. Serve.

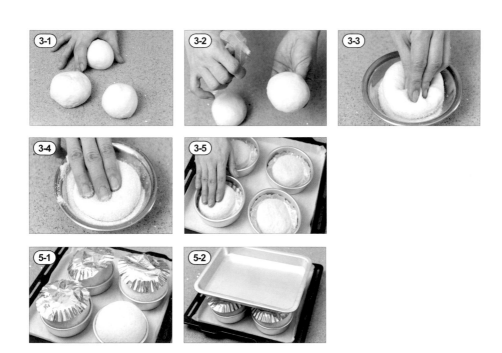

 小提醒 Tips

若沒有玉米粉，可用高筋麵粉或小麥胚芽代替。
If there is no corn flour, high gluten flour or wheat germ could be used instead.

日式照燒雞腿麵包
Japanese Teriyaki Chicken Bun

材料　Ingredients

高筋麵粉　250克/g
high gluten flour

酥油　20克/g
shortening

細鹽　5克/g
fine salt

奶粉　3克/g
milk powder

溫水　165毫升/ml
warm water

砂糖　15克/g
sugar

即溶酵母　4克/g
instant yeast

餡料　Filling

照燒雞腿餡料
（製法請參閱p.24）
（Refer to p.24 for the making
of teriyaki chicken filling）

表面　Topping

全蛋液（刷表面）　適量
some egg mixture for glazing

做法　Method

1. 詳細的基本麵包搓揉方法及第一次發酵方法請參閱 p.27 。

2. 將麵糰平均分割成 6 份，拍摺成長形後，讓麵糰鬆弛 15-20 分鐘。

3. 將麵糰按扁，由上向下捲，捲至尾部時將麵糰拉薄，繼續捲並將收口向下壓實，搓揉成橄欖形。麵包表面刷上少許全蛋液，沾上適量的白芝麻，放在烤盤上。進行最後發酵。（最後發酵方法請參閱 p.35）

4. 待麵糰發酵至兩倍大，放入已預熱的烤箱，用 180℃ 烤 15 分鐘即可。

1. For basic method of bread kneading and first fermentation, refer to p.27.

2. Divide the dough into 6 portions equally, roll into rods, let the dough rest for 15-20 minutes.

3. Press the dough to flatten. Roll the dough from the top to the bottom. Make the tip thinner and press firmly. Continue to roll. Invert the dough, place the openings at the bottom and press firmly. Shape into an olive. Brush the surface with egg mixture. Coat with sesame seeds. Place the doughs onto a baking tray for final fermentation. (Refer to method of final fermentation on p.35)

4. Wait until the dough rises 2 times in volume. Bake in a preheated oven at 180°Cfor 15 minutes. Serve.

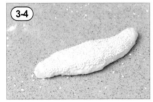

 小提醒 Tips

麵包的款式及形狀可隨個人喜好，若不喜歡做長形，可以改成圓形的款式。
The style and shape of the bun vary according to personal preferences, change to circular shape if olive shape is not preferred.

熱狗堡和豬扒包
Hot Dog And Pork Chop Bun

材料　Ingredients

高筋麵粉　250克/g
high gluten flour

酥油　20克/g
shortening

奶粉　3克/g
milk powder

細鹽　5克/g
fine salt

溫水　165毫升/ml
warm water

砂糖　15克/g
sugar

即溶酵母　4克/g
instant yeast

餡料　Filling

豬排（已煎熟）　3片
pork chop (pan-fried)

熱狗（已煎熟）　3條
sausages (pan-fried)

餡料　Filling

沙拉醬　適量
some salad dressing

生菜　適量
some lettuce

番茄　適量
some tomato

小黃瓜　適量
some cucumber

表面　Topping

全蛋液（刷表面）　適量
some egg mixture for glazing

做法　Method

1. 詳細的基本麵包搓揉方法及第一次發酵方法請參閱 p.27 。
2. 將麵糰平均分割成 6 份，分別拍摺成長形及搓揉成圓形後，讓麵糰鬆弛 15-20 分鐘。
3. 將三份長形麵糰按扁，由上向下捲，捲至尾部時將麵糰拉薄，繼續捲並將收口向下壓實，搓揉成橄欖形。另三個圓形再次搓圓並將空氣排出。各麵糰表面刷上少許全蛋液，沾上適量白芝麻，放在烤盤上，進行最後發酵。（最後發酵方法請參閱 p.35）
4. 待麵糰發酵至兩倍大，放入已預熱的烤箱，用 180℃烤 15 分鐘即可。
5. 麵包出爐放涼一會，切開，分別加入已預先煎熟的豬排或熱狗及其他餡料即可。

1. For basic method of bread kneading and first fermentation, refer to p.27.
2. Divide the dough into 6 portions equally, roll into rods and balls respectively, let the dough rest for 15-20 minutes.
3. Press 3 dough in rods to flatten. Roll the dough from the top to the bottom. Make the tip thinner. Continue to roll and invert the dough and place the openings at the bottom. Shape into an olive. Roll another 3 balls and release gas. Brush the doughs with some egg mixture, coat with white sesame seeds. Place the doughs onto a baking tray for final fermentation. (Refer to method of final fermentation on p.35)
4. Wail until the dough rises 2 times in volume. Bake in a preheated oven at 180°C for 15 minutes.
5. Leave the buns to cool. Cut horizontally, sandwich pan-fried pork chop or sausages and other filling. Serve.

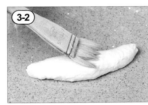

小提醒 Tips

麵包的大小和份量可按個人喜好，在進行麵糰分割前時要先決定好再分割。
The size of buns varies according to personal preferences, it should be decided when separating the dough.

日式咖哩牛肉麵包
Japanese Curry Beef Bun

材料　Ingredients

高筋麵粉　250克/g
high gluten flour

無鹽奶油（或酥油）
15克/g
unsalted butter
(or shortening)

奶粉　8克/g
milk powder

細鹽　2克/g
fine salt

溫水　135毫升/ml
warm water

砂糖　30克/g
sugar

雞蛋　25克/g
eggs

即溶酵母　4克/g
instant yeast

餡料　Filling

日式咖哩牛肉餡料
（製法請參考p.25）
（Refer to p.25 for the method of
making Japanese-style filling）

表面　Topping

全蛋液（刷表面）　適量
egg mixture for glazing

麵包粉（表面）　適量
some breadcrmbs for topping

做法 Method

1. 詳細的基本麵包搓揉方法及第一次發酵方法請參閱 p.27。

2. 將麵糰平均分割成 8 份，滾圓或拍摺成長形後，讓麵糰鬆弛 15-20 分鐘。

3. 將麵糰按扁，包入餡料，黏緊收口，麵糰兩面噴上少許清水，然後沾上適量的麵包粉，排放在烤盤上，進行最後發酵。（最後發酵方法請參閱 p.35）

4. 待麵糰發酵至兩倍大，放入已預熱的烤箱，用 180℃烤 15 分鐘即可。

1. For basic method of bread kneading and first fermentation, refer to p.27.

2. Divide the dough into 8 portions equally, roll into balls or rods, let the dough rest for 15-20 minutes.

3. Press the dough to flatten and add filling in the centre. Spray water on both sides of the dough. Coat with some breadcrumbs. Place the doughs onto a baking tray for final fermentation. (Refer to p.35 for method of final fermentation)

4. Wait until the dough rises 2 times in volume. Bake in a preheated oven at 180°C for 15 minutes. Serve.

 小提醒 Tips

若想品嘗原有脆口的風味，可以將麵包改用炸的方法代替烤的。或在麵包粉上刷上一層油再烤。

This recipe is improved for the sake of health. If you want to keep the original flavor of crunchy bun, deep-fry it instead of baking. Or brush oil onto the breadcrumbs before baking.

玩偶麵包
Doll Bun

材料　Ingredients

高筋麵粉　250克/g
high gluten flour

無鹽奶油　15克/g
unsalted butter

奶粉　8克/g
milk powder

細鹽　2克/g
fine salt

溫水　135毫升/ml
warm water

砂糖　30克/g
sugar

雞蛋　25克/g
eggs

即溶酵母　4克/g
instant yeast

表面裝飾　Topping Decoration

全蛋液（刷表面）　適量
egg mixture for glazing

黑豆（玩偶的眼睛）　適量
some black beans (doll eyes)

做法　Method

1. 詳細的基本麵包搓揉方法及第一次發酵方法請參閱 p.27。
2. 將麵糰平均分割成 9 份，滾圓後，讓麵糰鬆弛 15-20 分鐘。
3. 將麵糰按扁，並塑形。（另參考各款造型方法）
4. 將麵包排放在烤盤上，進行最後發酵。（最後發酵方法請參閱 p.37）
5. 待麵糰發酵至兩倍大，輕刷上全蛋液，放入已預熱的烤箱，用 180℃ 烤 15 分鐘即可。

1. For basic method of bread kneading and first fermentation, refer to p.27.
2. Divide the dough into 9 portions equally, roll into balls, let the dough rest for 15-20 minutes.
3. Press the dough to flatten and shape. (Refer to three different shapes)
4. Place the doughs onto a baking tray for final fermentation. (Refer to p.35 for method of final fermentation)
5. Wait until the dough rises 2 times in volume. Brush the surface with egg mixture. Bake in a preheated oven at 180°C for 15 minutes. Serve.

辮子女孩（3 個）　Girls with Twin Braids (3 servings)

將麵糰切出 1/3，按壓成長形及橫放，然後左右兩邊切成 3 條編成辮子，中間切 3 條做頭髮。另 2/3 麵糰揉圓做頭（如想加入餡料，可以自行加入）。將辮子髮型放在女孩頭上壓緊，再加上兩粒黑豆做眼睛即可。

Cut 1/3 from the dough, press into a rod and place horizontally, cut into 3 strips to form a braid from both left and right sides, cut into 3 strips as hairs in the middle of the head. Roll 2/3 of dough into a ball as a head. (Add filling if desired.) Place the braided hairs onto the girl's head and press firmly. Add 2 black beans as eyes.

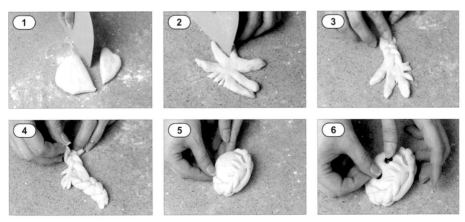

 小提醒 Tips

可以將出爐放涼後的麵包，用巧克力醬畫上眼睛或嘴巴。黑豆也可使用葡萄乾代替。
Leave the bread to cool and paint the eyes or mouth with chocolate sauce. Or use dried rasins instead of black beans.

魷魚麵包（3 個） Pig Head Bun (3 servings)

將麵糰切出 1/3，按壓成長形，然後再切成 1 份鼻子和 2 份耳朵。另 2/3 麵糰揉圓做頭（如想加入餡料，可以自行加入）。將兩隻耳朵放在豬頭上，再將另一塊鼻子麵糰，中間切兩條直條做成鼻孔，按壓在豬頭上，再加上兩粒黑豆做眼睛即可。

Cut 1/3 from the dough, roll into a rod. Divide into 1 nose and 2 ears. Roll 2/3 of dough into a ball as a head. (Add filling if desired.) Put 2 ears onto the pig's head. Cut 2 lines onto the piece of nose in the middle to make the nostrils. Press firmly onto the head. Add 2 black beans as eyes.

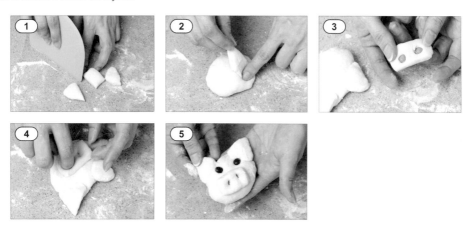

豬頭麵包（3 個） Squid Bun (3 servings)

1. 將麵糰切出 1/3，按壓成長形，然後切成 3 條，一端不要完全切斷。

2. 另 2/3 麵糰壓扁成三角形。將已切的長條麵糰放在三角形中央（如想加入餡料，可以自行加入）。將左右兩邊麵糰向內摺疊及壓實，再將麵糰翻轉，麵糰底部略為翻起成帽子，加上黑豆成為眼睛。

1. Cut 1/3 from the dough, roll into a rod. Cut the dough into 3 strips but do not cut off.

2. Press another 2/3 of dough into a triangle. Put the stripped dough onto the centre of the triangle. (Add filling if desired.) Fold the dough from both sides inwards and press firmly. Invert the dough and shape well, slightly turn up the bottom of the dough into a hat. Add black beans as the eyes.

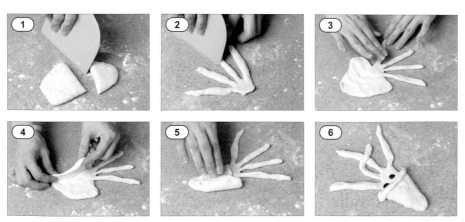

雞尾麵包
Cocktail Bun

湯種製法 The Method of Making Utane starter

湯種製作過程請參閱p.32
The method of making Utane starter, refer to p.32

1份湯種
1 Portion of Utane starter

高筋麵粉13克 +清水65毫升 → 製成品約有67克
13g high gluten flour + 65ml water → about 67g Utane starter

主麵糰（1份）
1 Portion of Main Dough

高筋麵粉　250克/g
high gluten flour

無鹽奶油　20克/g
unsalted butter

奶粉　9克/g
milk powder

細鹽　3克/g
fine salt

溫水　90毫升/ml
warm water

砂糖　35克/g
sugar

雞蛋　25克/g
egg

即用酵母　6克/g
instant yeast

湯種1份　67克/g
1 portion of Utane starter

餡料　Filling

雞尾麵包餡（製法請參照 p.24）
Cocktail filling (refer to p.24)

表面裝飾 Topping Decoration

墨西哥餡　適量（製法請參照 p.24）
some Mexican filling (refer to p.24)

白芝麻　適量
some white sesame seeds

全蛋液（刷表面）　適量
some egg mixture for glazin

做法　Method

1. 詳細的基本麵包搓揉方法及第一次發酵方法請參閱 p.27。

2. 將麵糰平均分割成 8 份，拍摺成長形後，讓麵糰鬆弛 15-20 分鐘。

3. 將麵糰按扁並輕輕拉長，將下方 1/3 部分麵糰摺向中央，餡料放在已摺疊的麵糰上，再將上方向下覆蓋餡料，然後輕壓收口並用手慢慢將麵糰向左右兩邊拉長，再按壓。最後將麵糰反轉收口，向底再壓實。

4. 將麵包排放在烤盤上，進行最後發酵。（最後發酵方法請參閱 p.35）

5. 待麵糰發酵至兩倍大，麵包表面刷上全蛋液及擠上兩條墨西哥餡料，然後灑上少許白芝麻，放入已預熱的烤箱，用 180℃烤 15 分鐘即可。

1. For basic method of bread kneading and first fermentation, refer to p.27.

2. Divide the dough into 8 portions equally, roll into rods, let the dough rest for 15-20 minutes.

3. Press the dough to flatten and stretch slightly, fold the lower 1/3 of the dough towards the centre. Put filling onto the folded dough, fold the upper part of the dough downwards to cover the filling. Press the opening slightly and stretch the dough on both sides slightly. Invert the dough and place the openings at the bottom and press firmly.

4. Place the doughs onto a baking tray for final fermentation. (Refer to p.35 for the method of final fermentation)

5. Wait until the dough rises 2 times in volume, brush the surface with egg mixture, pipe 2 strips of Mexican filling and sprinkle with some white sesame seeds. Bake in a preheated oven at 180°C for 15 minutes. Serve.

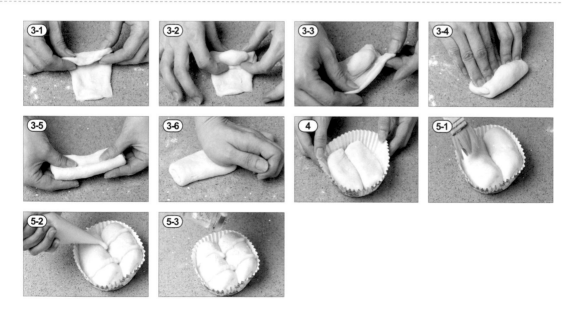

做雞尾麵包，最好與墨西哥麵包一起做，因為雞尾麵包上的兩條線是用墨西哥餡料，份量太少的話不易製作。

When making Cocktail Bun, it is better to make Mexican Bun at the same time, because the the Mexican filling is used for the 2 decorative strips for Cocktail Bun. It is hard to make the Mexican filling if the portion is too small.

墨西哥麵包
Mexican Bun

湯種製法 The Method of Making Utane starter

湯種製作過程請參閱p.32
The method of making Utane starter, refer to p.32

1份湯種
1 Portion of Utane starter

高筋麵粉13克 +清水65毫升
→ 製成品約有67克
13g high gluten flour + 65ml water → about 67g Utane starter

主麵糰（1份）
1 Portion of Main Dough

高筋麵粉　250克/g
high gluten flour

無鹽奶油　20克/g
unsalted butter

奶粉　9克/g
milk powder

細鹽　3克/g
fine salt

溫水　90毫升/ml
warm water

砂糖　35克/g
sugar

雞蛋　25克/g
egg

即用酵母　6克/g
instant yeast

湯種1份　67克/g
1 portion of Utane starter

餡料　Filling

墨西哥奶酥餡
（請參照 p.26 的製法）
Mexico filling (refer to p.26 for the method)

做法　Method

1. 詳細的基本麵包搓揉方法及第一次發酵方法請參閱 p.27 。
2. 將麵糰平均分割成 8 份，滾圓後，讓麵糰鬆弛 15-20 分鐘。
3. 將麵糰再次揉圓並將空氣排出，用手輕輕按壓麵糰。
4. 將麵包放在烤盤或麵包紙模上，進行最後發酵。（最後發酵方法請參閱 p.35）
5. 待麵糰發酵至兩倍大，在麵包表面擠上墨西哥餡，放入已預熱的烤箱，用 180℃烤 15 分鐘即可。

1. For basic method of bread kneading and first fermentation, refer to p.27.
2. Divide the dough into 8 portions equally, roll into balls, let the dough rest for 15-20 minutes.
3. Roll the dough into balls again, release gas, press the dough to flatten slightly.
4. Place the doughs onto a baking tray or onto a paper mold for final fermentation. (Refer to p.35 for the method of final fermentation)
5. Wait until the dough rises 2 times in volume, pipe Mexican filling onto the surface. Bake in a preheated oven at 180°C for 15 minutes. Serve.

小提醒 Tips

墨西哥麵包可以加入個人喜愛的餡料，如奶黃、紅豆餡等。
Other filling like egg custard or red bean paste could be added to Mexican Bun according to personal preference.

菠蘿麵包
Pineapple Bun

湯種製法 The Method of Making Utane starter

湯種製作過程請參閱p.32
The method of making Utane starter, refer to p.32

1份湯種
1 Portion of Utane starter

高筋麵粉13克 +清水65毫升
→ 製成品約有67克
13g high gluten flour + 65ml water → about 67g Utane starter

主麵糰（1份）
1 Portion of Main Dough

高筋麵粉　250克/g
high gluten flour

無鹽奶油　20克/g
unsalted butter

奶粉　9克/g
milk powder

細鹽　3克/g
fine salt

溫水　90毫升/ml
warm water

砂糖　35克/g
sugar

雞蛋　25克/g
egg

即用酵母　6克/g
instant yeast

湯種1份　67克/g
1 portion of Utane starter

表面裝飾　Topping Decoration

菠蘿麵包皮（做法請參閱p.24）
Pineapple topping
（Refer to p.24 for the method）

全蛋液（刷表面）　適量
some egg mixture for glazing

做法　Method

1. 詳細的基本麵包搓揉方法及第一次發酵方法請參閱 p.27 。
2. 將麵糰平均分割成 8 份，滾圓後，讓麵糰鬆弛 15-20 分鐘。
3. 將麵糰再次揉圓並將空氣排出。
4. 排放在烤盤上，進行最後發酵。（最後發酵方法請參閱 p.35）
5. 用一張保鮮膜對摺，將一份菠蘿皮麵包放在保鮮膜中央，用手指向四周推開成圓形的菠蘿皮。
6. 待麵糰發酵至兩倍大，每個麵糰放上一份菠蘿皮，表面輕刷全蛋液，放入已預熱的烤箱，用 180℃烤 15 分鐘即可。

1. For basic method of bread kneading and first fermentation, refer to p.27.
2. Divide the dough into 8 portions equally, roll into balls, let the dough rest for 15-20 minutes.
3. Roll the dough to balls again, release gas.
4. Place the doughs onto a baking tray for final fermentation. (Refer to p.35 for the method of final fermentation)
5. Fold a sheet of plastic wrap in halves, place a portion of pineapple topping onto the centre of plastic wrap, spread the topping outwards by fingers.
6. Wait until the dough rises 2 times in volume, top each dough with a sheet of pineapple topping. Brush the topping with egg mixture. Bake in a preheated oven at 180°C for 15 minutes. Serve.

小提醒 Tips

菠蘿麵包皮可以預先製作多份並分開包裝，放入冰箱冷凍，使用時預先改放冷藏格回軟，即可使用，可以存放一個月。

Pineapple topping could be pre-made and packaged separately in advance and store in the freezer for one month. Place the topping in refrigerated grid to soften before use.

法式軟麵包
French Soft Bun

湯種製法 The Method of Making Utane starter

湯種製作過程請參閱p.32
The method of making Utane starter, refer to p.32

1份湯種
1 Portion of Utane starter

高筋麵粉13克 +清水65毫升 → 製成品約有67克
13g high gluten flour + 65ml water → about 67g Utane starter

主麵糰（1份）
1 Portion of Main Dough

高筋麵粉　250克/g
high gluten flour

無鹽奶油　20克/g
unsalted butter

奶粉　9克/g
milk powder

細鹽　3克/g
fine salt

溫水　90毫升/ml
warm water

砂糖　35克/g
sugar

雞蛋　25克/g
egg

即用酵母　6克/g
instant yeast

湯種1份　67克/g
1 portion of Utane starter

表面裝飾 Topping Decoration

黑色櫻桃籽或高筋麵粉
black cherry seeds or high gluten flour

做法　Method

1. 詳細的基本麵包搓揉方法及第一次發酵方法請參閱 p.27 。
2. 將麵糰平均分割成 8 份，拍摺成長形後，讓麵糰鬆弛 15-20 分鐘。
3. 將麵糰按扁，由上向下捲，捲至尾部時將麵糰拉薄，繼續捲並將收口向下壓實，揉成橄欖形，表面噴上少許水，灑上適量的黑色櫻桃籽。
4. 將麵包排放在烤盤上，進行最後發酵。（最後發酵方法請參閱 p.35）
5. 待麵糰發酵至兩倍大，麵包表面噴上清水，然後灑上少許高筋麵粉，再用刀劃幾下，立即放入已預熱的烤箱，用 180℃烤 15 分鐘即可。

1. For basic method of bread kneading and first fermentation, refer to p.27.
2. Divide the dough into 8 portions equally, roll into rods, let the dough rest for 15-20 minutes.
3. Press the dough to flatten slightly. Roll the dough from the top to the bottom. Make the tip thinner and press firmly. Continue to roll, invert the dough and place the openings at the bottom. Shape into an olive. Spray with some water and then sprinkle with some black cherry seeds.
4. Place the doughs onto a baking tray for final fermentation. (Refer to p.35 for the method of final fermentation)
5. Wait until the dough rises 2 times in volume. Spray with some water and then some high gluten flour. Slit several lines with a blade. Bake in a preheated oven at 180°C for 15 minutes. Serve.

 小提醒 Tips

法式軟包還可以搓揉成圓形，待發酵完成後，烤製前才在麵包表面噴少許水及灑上高筋麵粉作裝飾。
French soft bun can also be shaped in round, when fermentation is completed, spray some water and sprinkle with high gluten flour for decoration before baking.

北海道麵包
Hokkaido Bun

湯種製法 The Method of Making Utane starter

湯種製作過程請參閱p.32
The method of making Utane starter, refer to p.32

1份湯種
1 Portion of Utane starter

高筋麵粉13克 +清水65毫升
→ 製成品約有67克
13g high gluten flour + 65ml water → about 67g Utane starter

主麵糰（1份）
1 Portion of Main Dough

高筋麵粉　250克/g
high gluten flour

無鹽奶油　20克/g
unsalted butter

奶粉　9克/g
milk powder

細鹽　3克/g
fine salt

溫水　90毫升/ml
warm water

砂糖　35克/g
sugar

雞蛋　25克/g
egg

即溶酵母　6克/g
instant yeast

湯種1份　67克/g
1 portion of Utane starter

餡料 Filling

卡士達起士餡（請參閱 p.25）
filling of custard cream (Refer to p.25)

表面裝飾 Topping Decoration

北海道麵包餡料（請參閱p.24）
refer to Hokkaido Bun topping (refer to p.24)

做法　Method

1. 詳細的基本麵包搓揉方法及第一次發酵方法請參閱 p.27 。

2. 將麵糰平均分割成 8 份，揉圓後，讓麵糰鬆弛 15-20 分鐘。

3. 將麵糰按扁，包入 1 湯匙餡料，黏緊收口，將麵糰翻轉，收口向下，用手輕輕按壓麵糰，然後放入模內。

4. 將麵包排放在烤盤上，進行最後發酵。（最後發酵方法請參閱 p.35）

5. 待麵糰發酵至兩倍大，擠上約 8 成滿的表面餡料，放入已預熱的烤箱，用 180℃烤 15 分鐘即可。

1. For basic method of bread kneading and first fermentation, refer to p.27.

2. Divide the dough into 8 portions equally, roll into balls, let the dough rest for 15-20 minutes.

3. Press the dough to flatten slightly. Add 1 tbsp of filling.Press the opening firmly. Invert the dough and place the openings at the bottom. Press the dough to flatten slightly. Put into a mold.

4. Place the doughs onto a baking tray for final fermentation. (Refer to p.35 for the method of final fermentation)

5. Wait until the dough rises 2 times in volume. Pipe with filling until 80% full. Bake in a preheated oven at 180°C for 15 minutes. Serve.

若想麵包有香濃牛奶香味，可用北海道牛奶代替溫水。
If thick and smooth milk flavor is desired, use Hokkaido milk instead of warm water.

巧克力卡士達麵包
Chocolate Custard Bun

湯種製法 The Method of Making Utane starter

湯種製作過程請參閱p.32
The method of making Utane starter, refer to p.32

1份湯種
1 Portion of Utane starter

高筋麵粉13克 +清水65毫升
→ 製成品約有67克
13g high gluten flour + 65ml water → about 67g Utane starter

主麵糰（1份）
1 Portion of Main Dough

高筋麵粉　250克/g
high gluten flour

無鹽奶油　20克/g
unsalted butter

奶粉　9克/g
milk powder

細鹽　3克/g
fine salt

溫水　90毫升/ml
warm water

砂糖　35克/g
sugar

雞蛋　25克/g
egg

即溶酵母　6克/g
instant yeast

湯種1份　67克/g
1 portion of Utane starter

餡料 Filling

巧克力卡士達餡（請參閱 p.27）
filling of chocolate custard
(refer to p.27)

表面裝飾 Topping Decoration

巧克力卡士達麵包表面（請參閱 p.27）
topping of chocolate custard
(refer to p.27)

做法　Method

1. 詳細的基本麵包搓揉方法及第一次發酵方法請參閱 p.27。

2. 將麵糰平均分割成 8 份，揉圓後，讓麵糰鬆弛 15-20 分鐘。

3. 將麵糰按扁，包入 1 湯匙餡料，黏緊收口，將麵糰翻轉收口向下，用手輕輕按壓麵糰，然後放入模內。

4. 將包排放在烤盤上，進行最後發酵。（最後發酵方法請參閱 p.35）

5. 待麵糰發酵至兩倍大，擠上約 8 成滿的表面餡料，放入已預熱的烤箱，用 180℃ 烤 15 分鐘即可。

1. For basic method of bread kneading and first fermentation, refer to p.27.

2. Divide the dough into 8 portions equally, roll into balls, let the dough rest for 15-20 minutes.

3. Press the dough to flatten slightly. Add 1 tbsp of filling.Press the opening firmly. Invert the dough and place the openings at the bottom. Press the dough to flatten slightly. Put into a mold.

4. Place the doughs onto a baking tray for final fermentation. (Refer to p.35 for the method of final fermentation)

5. Wait until the dough rises 2 times in volume. Pipe with filling until 80% full. Bake in a preheated oven at 180°C for 15 minutes. Serve.

可以在麵包表面灑上少許白芝麻，使外表更加吸引人。
To make the appearance more attractive, sprinkle some white sesame seeds on the surface.

鮮奶麵包
Fresh Milk Row Bread

湯種製法 The Method of Making Utane starter

湯種製作過程請參閱p.32
The method of making Utane starter, refer to p.32

1份湯種
1 Portion of Utane starter

高筋麵粉13克 +清水65毫升 → 製成品約有67克
13g high gluten flour + 65ml water → about 67g Utane starter

主麵糰（1份）
1 Portion of Main Dough

高筋麵粉 200克/g
high gluten flour

奶油 20克/g
butter

奶粉 8克/g
milk powder

鹽 2克/g
fine salt

溫鮮奶 75毫升/ml
warm milk

砂糖 35克/g
sugar

雞蛋 20克/g
egg

酵母 5克/g
instant yeast

湯種 55克/g
1 portion of Utane starter

表面 Topping Decoration

全蛋液（刷表面） 適量
some egg mixture for glazing

做法　Method

1. 詳細的基本麵包搓揉方法及第一次發酵方法請參閱 p.27 。
2. 將麵糰平均分割成 8 份，拍摺成長形後，讓麵糰鬆弛 15-20 分鐘。
3. 將麵糰按扁，由上向下捲，捲至尾部時將麵糰拉薄，繼續捲並將收口向下壓實。
4. 預先用鋁箔紙摺一個約 8 吋 ×8 吋的方形鋁箔紙模，將麵糰平均排放在模型上。
5. 將麵糰放在烤盤上，進行最後發酵。（最後發酵方法請參閱 p.35 ）
6. 待麵糰發酵至兩倍大，在表面輕刷全蛋液，放入已預熱的烤箱，用 170℃烤 15 分鐘即可。

1. For basic method of bread kneading and first fermentation, refer to p.27.
2. Divide the dough into 8 portions equally, roll into rods, let the dough rest for 15-20 minutes.
3. Press the dough to flatten slightly. Roll the dough from the top to the bottom. Make the tip thinner and press firmly. Invert the dough and place the openings at the bottom.
4. Fold a sheet of alulinium foil and make a square mold measures 8 inch x 8 inch. Place the doughs onto the mold evenly.
5. Place the doughs onto a baking tray for final fermentation. (Refer to p.35 for the method of final fermentation)
6. Wait until the dough rises 2 times in volume. Brush the Brush the surface with egg mixture. Bake in a preheated oven at 170°C for 15 minutes. Serve.

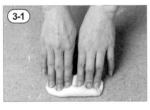 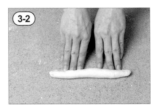 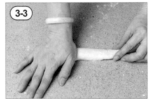

 小提醒 Tips

1. 用北海道牛奶或將部分鮮奶換成動物性鮮奶油，可使麵包更香濃。
2. 若沒有烤盤將麵包定形，可以用鋁箔紙摺成方形來固定麵包形狀。
1. To make the bread more creamy, use Hokkaido milk or whipping cream instead of milk.
2. If you don't have a baking tray to shape the bread, fold a mold with aluminum foil.

南瓜籽麵包捲
Pumpkin Seed Roll

1份湯種
1 Portion of Utane starter

高筋麵粉13克 +清水65毫升
→ 製成品約有67克
13g highgluten flour + 65ml
water → about 67g Utane
starter

主麵糰（1份）
1 Portion of Main Dough

高筋麵粉　250克/g
high gluten flour

奶油　15克/g
unsalted butter

奶粉　6克/g
milk powder

鹽　2克/g
fine salt

砂糖　40克/g
sugar

雞蛋　25克/g
egg

溫水　20毫升/ml
warm water

酵母　6克/g
instant yeast

南瓜泥　80克/g
pumpkin puree

湯種1份　67克/g
1 portion of Utane starter

表面裝飾　Topping Decoration

南瓜籽　適量
some pumpkin seeds

全蛋液（刷表面）　適量
some egg mixture for glazing

表面　Topping Decoration

全蛋液（刷表面）　適量
some egg mixture for glazing

份量　Servings

2條三辮包（半份麵糰）及4個小圓包
（半份麵糰）（2 rolls braided bread
with 1/2 dough and 4 round buns with
1/2 dough）

做法　Method

1. 將半份麵糰平均分為 2 份，拍摺成長形後，讓麵糰鬆弛 15-20 分鐘。
2. 將每份麵糰按扁並輕輕拉長，將麵糰切成 3 條，一邊連著不要切斷。
3. 將麵糰左右交叉編成辮子狀，收口壓實，然後捲成圓形放入紙模內。
4. 將麵糰放在烤盤上，進行最後發酵。
5. 待麵糰發酵至兩倍大，輕刷上全蛋液及灑上南瓜籽，放入已預熱的烤箱，用 180℃ 烤 15 分鐘即可。

1. Divide the dough into 2 portions equally, roll into rods, let the dough rest for 15-20 minutes.
2. Press the dough to flatten and stretch slightly. Cut the dough into 3 strips but do not cut off.
3. Cross weave into braids like shape, press the opening. Shape like a circle and put into a round mold.
4. Place the doughs onto a baking tray for final fermentation. (Refer to p.35 for the method of final fermentation)
5. Wait until the dough rises 2 times in volume. Brush the surface with egg mixture. Top with pumpkin seeds. Bake in a preheated oven at 180°C for 15 minutes. Serve.

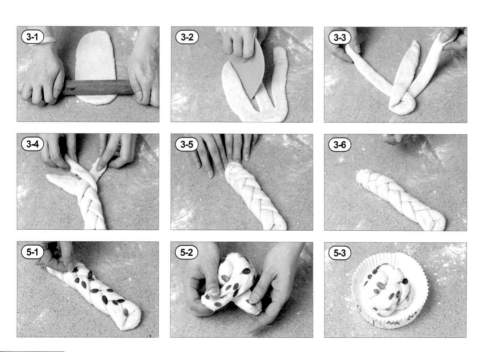

小提醒 Tips

1. 若將半份麵糰改做小圓麵包，應在麵糰分割時分成4份並搓成圓形狀，然後鬆弛。
2. 南瓜泥的水分在蒸壓後可能加多減少，搓揉麵糰時，若發覺水分不足，應自行添加。

1. If you want to make round buns by half of the dough, divide the dough into 4 portions and roll into balls and then rest.
2. After steaming pumpkin puree, the proportion of water may vary. Adjust the amount of water when kneading the dough.

葵花亞麻籽南瓜卡士達麵包
Sunflower Flaxseed and Pumpkin Custard Bun

1份湯種
1 Portion of Utane starter

高筋麵粉13克 +清水 65毫升→ 製成品約有67克
13g highgluten flour + 65ml water → about 67g Utane starter

主麵糰（1份）
1 Portion of Main Dough

高筋麵粉　250克/g
high gluten flour

奶油　15克/g
butter

奶粉　6克/g
milk powder

鹽　2克/g
fine salt

雞蛋　25克/g
egg

砂糖　40克/g
sugar

溫水　20毫升/ml
warm water

酵母　6克/g
instant yeast

湯種1份　67克/g
1 portion of Utane starter

南瓜泥　80克/g
pumpkin puree

南瓜卡士達餡料
Pumpkin Custard Filling

南瓜（製法請參閱 p.28）　200克/g
flesh pumpkin (refer to p.28)

表面裝飾 Topping Decoration

葵花籽及亞麻籽 適量
some sunflower seeds and flaxseeds

全蛋液（刷表面）　適量
some egg mixture for glazing

做法　Method

1. 詳細的基本麵包搓揉方法及第一次發酵方法請參閱 p.27 。

2. 將麵糰平均分割成 8 份，滾圓後，讓麵糰鬆弛 15-20 分鐘。

3. 將麵糰按扁，包入適量餡料，黏緊收口，將麵糰反轉收口向下，用手輕輕按壓麵糰。

4. 將麵包排放在烤盤上，進行最後發酵。（參考最後發酵方法 p.35）

5. 待麵糰發酵至兩倍大，在表面輕刷上全蛋液，並灑上葵花籽及亞麻籽，放入已預熱的烤箱，用 180℃烤 15 分鐘即可。

1. For basic method of bread kneading and first fermentation, refer to p.27.

2. Divide the dough into 8 portions equally, roll into balls, let the dough rest for 15-20 minutes.

3. Press the dough to flatten and add some filling. Press the opening. Invert the dough and place the openings at the bottom. Press the dough to flatten slightly.

4. Place the doughs onto a baking tray for final fermentation. (Refer to p.35 for the method of final fermentation)

5. Wait until the dough rises 2 times in volume, brush the surface with egg mixture, sprinkle with sunflower seeds and flaxseeds. Bake in a preheated oven at 180□for 15 minutes. Serve.

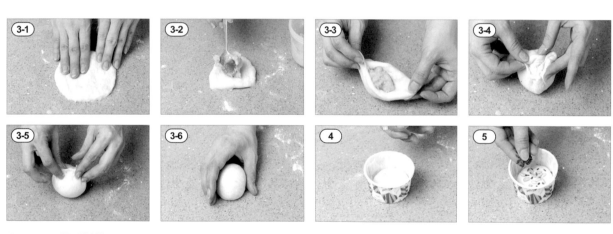

註：可用長形紙模　Use square mold

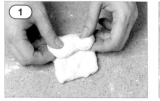

 小提醒 Tips

南瓜泥的水分在蒸壓後可能有增多減少，搓揉麵糰時，若發覺水分不足，應自行添加。
After steaming pumpkin puree, the proportion of water may vary. Adjust the amount of water when kneading the dough.

葡萄乾大麥麵包 / 葡萄乾小麥麵包
Raisin Whole Meal Bread /
Mini Raisin Whole Meal Bread

材料　Ingredients

高筋麵粉　250克/g
high gluten flour

奶油　20克/g
butter

奶粉　8克/g
milk powder

麥芽精　5克/g
malt extract

鹽　2克/g
salt

溫水　125毫升/ml
warm water

雞蛋　25克/g
egg

即溶酵母　5克/g
instant yeast

黑糖　45克/g
brown sugar

葡萄乾　80克/g
raisins

餡料　Filling

葡萄乾　80克/g
raisins

蘭姆酒　1茶匙/tsp
rum

份量　Servings

中型提子包1個+4個細圓提子包
（1 medium raisin bread + 4 mini
round breads）

100

做法　Method

1. 詳細的基本麵包搓揉方法及第一次發酵方法請參閱 p.27 。
2. 葡萄乾用熱水略泡 1 分鐘，瀝乾水後加入蘭姆酒浸 1 晚或最少 1 小時。
3. 將葡萄乾加入麵糰內搓勻及平均分割成 5 份（半份麵糰做大麥麵包，另 4 份細麵糰做小麥葡萄乾麵包），分別拍摺成長形及搓揉成圓形後，讓麵糰鬆弛 15-20 分鐘。
4. 將麵糰按扁排氣，將麵糰由上向下摺，將收口壓實，搓成橄欖形。若做小麥麵包，要將麵糰再按壓排氣，才進行最後發酵。
5. 將麵包排放在烤盤上，進行最後發酵。（最後發酵方法請參閱 p.35）
6. 待麵糰發酵至兩倍大，在表面刷上全蛋液，放入已預熱的烤箱，用 180℃烤 15 分鐘即可。

1. For basic method of bread kneading and first fermentation, refer to p.27.
2. Soak raisins in hot water for 1 minute. Drain and soak in rum overnight or at least 1 hour.
3. Add raisins into the dough and knead well. Divide the dough into 5 portions equally. (Make Raisin Whole Meal Bread by 1/2 dough and 4 Mini Raisin Whole Meal Bread by another half) Roll into rods and balls respectively, let the dough rest for 15-20 minutes.
4. Press the dough to flatten and release gas. Roll the dough from the top to the bottom. Press the opening firmly. Shape into an olive. For Mini Raisin Whole Meal Bread, press to flatten and release gas again before final fermentation.
5. Place the doughs onto a baking tray for final fermentation. (Refer to p.35 for method of final fermentation)
6. Wait until the dough rises 2 times in volume. brush the surface with egg mixture. Bake in a preheated oven at 180°C for 15 minutes. Serve.

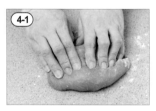
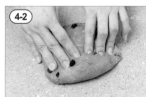

小提醒 Tips

1. 葡萄乾亦可以待麵糰平均分割後才加入揉勻。
2. 麵糰鬆弛後的排氣動作要認真做，盡量將麵糰按壓排氣，否則烤後麵糰表面會形成皺紋。

1. Raisins could be added and kneaded after the dough is divided.
2. The process of releasing gas after resting is important and should take more attention, try to press the dough and release, otherwise the surface of the dough will wrinkle after baking.

黑糖核桃麵包
Brown Sugar and Walnut Bread

材料　Ingredients

高筋麵粉　250克/g
high gluten flour

奶油　15克/g
butter

奶粉　8克/g
milk powder

鹽　2克/g
salt

溫水　130毫升/ml
warm water

雞蛋　25克/g
egg

即溶酵母　5克/g
instant yeast

黑糖　45克/g
brown sugar

已烘烤的核桃　60克/g
baked walnuts

表面　Topping

全蛋液（刷表面）　適量
some egg mixture for glazing

做法　Method

1. 詳細的基本麵包搓揉方法及第一次發酵方法請參閱 p.27 。
2. 將麵糰平均分割成 8 份，搓揉成圓形或橄欖形後，讓麵糰鬆弛 15-20 分鐘。
3. 將麵糰按扁，並核桃加入搓圓或摺成橄欖形，收口壓實。
4. 將麵包排放在烤盤上，進行最後發酵。（最後發酵方法請參閱 p.35）
5. 待麵糰發酵至兩倍大，在表面輕刷上全蛋液，放入已預熱的烤箱，用 180℃烤 15 分鐘即可。

1. For basic method of bread kneading and first fermentation, refer to p.27.
2. Divide the dough into 8 portions equally, roll into balls or olives, let the dough rest for 15-20 minutes.
3. Press the dough to flatten. Add walnuts and knead well, roll into balls or olives. Press the opening firmly.
4. Place the doughs onto a baking tray for final fermentation. (Refer to p.35 for method of final fermentation)
5. Wait until the dough rises 2 times in volume. brush the surface with egg mixture. Bake in a preheated oven at 180°C for 15 minutes. Serve.

 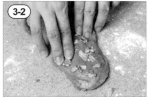  

 小提醒 Tips

核桃餡可以改用紅莓乾，只要預先用熱水沖洗紅莓乾，加入少許蘭姆酒浸一晚，吃時會更軟，香味更佳。
Dried cranberries can be used instead of walnuts for filling. Rinse dried cranberries in hot water, soak in rum overnight, they will be softer and the flavor will be better.

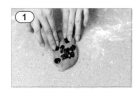  

農夫麵包
Farmer Bread

材料　Ingredients

高筋麵粉　150克/g
high gluten flour

全麥麵粉　80克/g
whole wheat flour

黑裸麥粉　15克/g
rye flour

燕麥　10克/g
oats

奶油　10克/g
butter

奶粉　5克/g
milk powder

鹽　4克/g
salt

溫水　170-175毫升/ml
warm water

砂糖　15克/g
sugar

即溶酵母　5克/g
instant yeast

表面　Topping

高筋麵粉　適量
some high gluten flour

做法　Method

1. 詳細的基本麵包搓揉方法及第一次發酵方法請參閱 p.27。
2. 將麵糰平均分割成 2 份，拍摺成長形後，讓麵糰鬆弛 15-20 分鐘。
3. 將麵糰按扁排氣，將麵糰由上向下拍摺壓後推成圓形。
4. 將麵糰排放在烤盤上，進行最後發酵。（參考最後發酵方法 P.35）
5. 待麵糰發酵至兩倍大，表面噴少許水，再灑上高筋麵粉，然後用刀片劃幾刀，放入已預熱的烤箱，並朝烤箱噴水數下。用 180℃烤 25-30 分鐘即可。

1. For basic method of bread kneading and first fermentation, refer to p.27.
2. Divide the dough into 2 portions equally, roll into rods, let the dough rest for 15-20 minutes.
3. Press the dough to flatten. Release gas. Roll from the top to the bottom, roll into balls.
4. Place the doughs onto a baking tray for final fermentation. (Refer to p.35 for method of final fermentation)
5. Wait until the dough rises 2 times in volume. Spray with some water, sprinkle with some high gluten flour. Make several slits with a blade. Place into a preheated oven and spray some water into the oven. Bake at 180°C for 25-30 minutes. Serve.

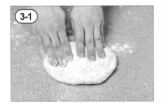

 小提醒 Tips

可以用蕎麥粉代替黑稞粉。
Buckwheat powder could be used instead of rye f lour.

粗麥麵包
Coarse Whole Meal Bread

材料　Ingredients

高筋麵粉　150克/g
high gluten flour

全麥麵粉　80克/g
whole wheat flour

奶油　20克/g
butter

燕麥片　20克/g
oatmeal

亞麻子　10克/g
linseeds

麥芽精　5克/g
malt extract

鹽　4克/g
salt

溫水　165毫升/ml
warm water

砂糖　20克/g
sugar

即溶酵母　4克/g
instant yeast

表面　Topping

亞麻籽（預先烤脆）及高筋麵粉
Flaxseeds (baked till crispy)
and high gluten flour

做法　Method

1. 詳細的基本麵包搓揉方法及第一次發酵方法請參閱 p.27。
2. 將麵糰平均分割成 2 份，拍摺成長形後，讓麵糰鬆弛 15-20 分鐘。
3. 將麵糰按扁並輕輕拉長，將下方 1/3 部分麵糰摺向中央，麵糰上方摺向中央，再捲向底部，將收口壓實，搓揉成橄欖形。
4. 將麵包排放在烤盤上，進行最後發酵。（最後發酵方法請參閱 p.35）
5. 待麵糰發酵至兩倍大，表面噴少許水，灑上亞麻籽及高筋麵粉，然後用刀片劃幾刀，放入已預熱的烤箱，朝烤箱噴水數下。用 180℃ 烤 25-30 分鐘即可。

1. For basic method of bread kneading and first fermentation, refer to p.27.
2. Divide the dough into 2 portions equally, roll into rods, let the dough rest for 15-20 minutes.
3. Press the dough to flatten and stretch slightly. Fold the lower 1/3 portion of the dough upwards to the centre. Fold the upper 1/3 portion towards the centre. Roll towards the bottom. Press the opening firmly. Roll into an olive.
4. Place the doughs onto a baking tray for final fermentation. (Refer to p.35 for method of final fermentation)
5. Wait until the dough rises 2 times in volume. Spray with some water. Sprinkle with flaxseeds and high gluten flour. Make several slits with a blade. Place into a preheated oven and spray some water into the oven. Bake at 180°C for 25-30 minutes. Serve.

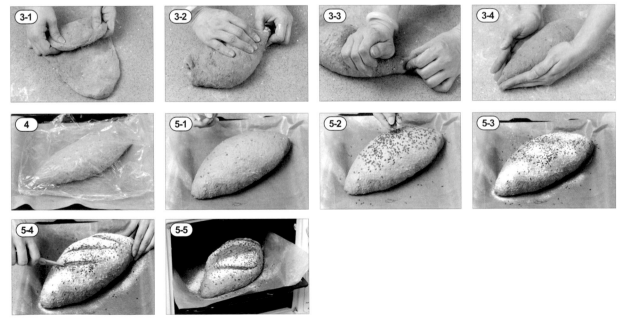

 小提醒 Tips

麵包未送入烤箱前，千萬別太早用刀劃開表面，要等到送入烤箱前才做此動作。
Do not make the slits too early, wait until it is ready to put into the oven.

燕麥麵包
Oatmeal Bread

材料　Ingredients

高筋麵粉　250克/g
high gluten flour

酥油　15克/g
shortening

奶粉　8克/g
milk powder

細鹽　2克/g
fine salt

溫水　135毫升/ml
warm water

砂糖　30克/g
sugar

雞蛋　25克/g
egg

即溶酵母　4克/g
instant yeast

表面　Topping Filling

燕麥片　適量
some oatmeal

做法　Method

1. 詳細的基本麵包搓揉方法及第一次發酵方法請參閱 p.27。
2. 將麵糰平均分割成 8 份，滾圓後，讓麵糰鬆弛 15-20 分鐘。
3. 將麵糰再次搓圓，將空氣排出，在麵糰表面噴少許水，然後沾上燕麥。
4. 將麵包排放在烤盤上，進行最後發酵。（參考最後發酵方法 P.35）
5. 待麵糰發酵至兩倍大，放入已預熱的烤箱，用 180℃烤 15 分鐘即可。

1. For basic method of bread kneading and first fermentation, refer to p.27.
2. Divide the dough into 8 portions equally, roll into balls, let the dough rest for 15-20 minutes.
3. Roll the dough into balls and release gas. Spray some water onto the surface and coat with oatmeal.
4. Place the doughs onto a baking tray for final fermentation. (Refer to p.35 for method of final fermentation)
5. Wait until the dough rises 2 times in volume. Bake in a preheated oven at 180°C for 15 minutes. Serve.

 小提醒 Tips

想吃得更健康，平常可用果醬塗在燕麥包上，或夾其他餡料來吃。
It is healthier to have Oatmeal Bread served with jam or other filling.

海苔肉鬆方形麵包
Seaweed And Floss Bun

麵包模　Bread Mold

4吋×6.5吋 長方形烤模
4 " X 6.5 " rectangular
baking mold

材料　Ingredients

高筋麵粉　250克/g
high gluten flour

酥油　15克/g
shortening

奶粉　8克/g
milk powder

細鹽　2克/g
fine salt

溫水　135毫升/ml
warm water

砂糖　30克/g
sugar

雞蛋　25克/g
egg

即溶酵母　4 克/g
instant yeast

餡料　Filling

豬肉鬆　30克/g
pork floss

海苔片　適量
some seaweed

馬自瑞拉起士　適量
some mozzarella

全蛋液（刷表面）　適量
some egg mixture for glazing

做法　Method

1. 詳細的基本麵包搓揉方法及第一次發酵方法請參閱 p.27 。
2. 將麵糰平均分割成 2 份，拍摺成長形，讓麵糰鬆弛 15-20 分鐘。
3. 將麵糰按扁及用木棍桿薄成 7 吋 X10 吋長形，放上豬肉鬆及海苔片，然後將麵糰由上向下捲起，收口略為拉薄並壓實。
4. 放入已刷油的模型內，進行最後發酵。（參考最後發酵方法 p.35）
5. 待麵糰發酵至 8 成，輕刷上全蛋液，灑上少許起士碎粒。放入已預熱的烤箱底層，用 180℃ 烤 15 分鐘，然後改放在中層，表面蓋上鋁箔紙再烤 10 分鐘。麵包出爐即脫模放涼即可。

1. For basic method of bread kneading and first fermentation, refer to p.27.
2. Divide the dough into 2 portions equally, roll into rods, let the dough rest for 15-20 minutes.
3. Press the dough to flatten and roll into rectangular shape of 7" x 10" by a rolling pin, arrange pork floss and seaweed onto the dough evenly. Roll the dough from the top to the bottom. Make the tip thinner and press firmly.
4. Place the dough into a greased mold for final fermentation. (Refer to method of final fermentation on p.35)
5. Wait until the dough rises to 80% full, brush the surface with egg mixture. Sprinkle with shredded mozzarella. Bake in the lower deck of a preheated oven at 180°C for 15 minutes. Change to the middle deck. Cover with a sheet of aluminium foil and bake for 10 more minutes. Leave to cool and remove mold. Serve.

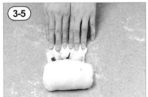

小提醒 Tips

灑起士碎粒時要留意放在麵包中央，盡量不要沾在烤模旁邊，否則麵包脫模時會黏住，不易脫模。
When sprinkling mozzarella onto the dough, make sure that it is in the centre but not onto the sides of the mold, otherwise it is difficult to remove the mold.

肉桂葡萄乾方形麵包
Cinnamon and Raisin Bread

麵包模　Bread Mold

4吋×6.5吋 長方形烤模
4" X 6.5" rectangular
baking mold

材料　Ingredients

高筋麵粉　250克/g
high gluten flour

無鹽奶油（或酥油）15克/g
unsalted butter
(or shortening)

奶粉　8克/g
milk powder

細鹽　2克/g
fine salt

溫水　135毫升/ml
warm water

砂糖　30克/g
sugar

雞蛋　25克/g
egg

即溶酵母　4克/g
instant yeast

餡料　　Filling

葡萄乾 30克/g
raisins

蘭姆酒　1茶匙/tsp
rum

肉桂粉　適量
some cinnamon powder

全蛋液（刷表面）　適量
some egg mixture for glazing

做法　Method

1. 詳細的基本麵包搓揉方法及第一次發酵方法請參閱 p.27。
2. 將麵糰平均分割成 2 份，拍摺成長形，讓麵糰鬆弛 15-20 分鐘。
3. 葡萄乾用熱水略泡 1 分鐘，瀝乾水後加入蘭姆酒泡 1 晚或最少 1 小時。
4. 將麵糰按扁及用木棍桿薄成 6 吋 X12 吋長形，將葡萄乾及肉桂粉平均放在麵糰上，麵糰由上向下捲起，收口略為拉薄並壓實。
5. 放入已刷油的模型內，進行最後發酵。（最後發酵方法請參閱 p.35）
6. 待麵糰發酵至 8 成，輕刷上全蛋液，放入已預熱的烤箱底層，用 180℃烤 15 分鐘，然後改放在中層，表面蓋上鋁箔紙再烤 10 分鐘。麵包出爐即脫模放涼。

1. For basic method of bread kneading and first fermentation, refer to p.27.
2. Divide the dough into 2 portions equally, roll into rods, let the dough rest for 15-20 minutes.
3. Soak raisins in hot water for 1 minute, drain and soak in rum overnight or at least 1 hour.
4. Press the dough to flatten and roll into rectangular shape of 6" x 12" by a rolling pin, arrange raisins and cinnamon powder onto the dough evenly. Roll the dough from the top to the bottom. Make the tip thinner and press firmly.
5. Place the doughs into a greased mold for final fermentation. (Refer to method of final fermentation on p.35)
6. Wait until the dough rises to 80 % full, brush the surface with egg mixture. Bake in the lower deck of a preheated oven at 180°C for 15 minutes. Change to the middle deck. Cover with a sheet of aluminium foil and bake for 10 more minutes. Leave to cool and remove mold. Serve.

小提醒 Tips

葡萄乾用蘭姆酒浸一晚，會更香甜及軟。
Raisins soaked overnight will be more fragrant and soft.

洋蔥火腿方形麵包
Onion and Ham Bun

麵包模　Bread Mold

4吋×6.5吋 長方形烤模
4" X 6.5 " rectangular
baking mold

材料　　Ingredients

高筋麵粉　250克/g
high gluten flour

酥油　20克/g
shortening

砂糖　15克/g
sugar

奶粉　3克/g
milk powder

溫水　160毫升/ml
warm water

細鹽　5克/g
fine salt

即溶酵母　4克/g
instant yeast

餡料　　Filling

洋蔥粒　60克/g
diced onion

火腿粒　80克/g
diced ham

細鹽　1/4茶匙/tsp
fine salt

表面　　Topping

沙拉醬及馬自瑞拉起士碎粒　適量
some salad dressing and chopped
mozzarella

全蛋液（刷表面）　適量
some egg mixture for glazing

做法　Method

1. 詳細的基本麵包搓揉方法及第一次發酵方法參閱 p.27。
2. 將麵糰平均分割成 2 份，拍摺成長形後，讓麵糰鬆弛 15-20 分鐘。
3. 洋葱用少許油炒至軟化，然後加入火腿及細鹽略炒，備用。
4. 將麵糰按扁，用木棍桿薄成 6 吋 X12 吋長形，將餡料平均鋪在麵糰上，輕輕壓實，兩邊麵糰向內摺 1 厘米，然後將麵糰由上向下捲起，收口略為拉薄並壓實。
5. 放入已刷油的模型內，進行最後發酵。（最後發酵方法請參閱 p.35）
6. 待麵糰發酵至 8 成，在表面輕刷上全蛋液，擠上適量沙拉及灑上少許起士碎粒，放入已預熱的烤箱底層，用 180℃ 烤 15 分鐘，然後改放在中層，表面蓋上鋁箔紙再烤 10 分鐘。麵包出爐即脫模放涼。

1. For basic method of bread kneading and first fermentation, refer to p.27.
2. Divide the dough into 2 portions equally, roll into rods, let the dough rest for 15-20 minutes.
3. Stir-fry diced onion by some oil till soft, add ham and fine salt and stir-fry for a while, dish up for later use.
4. Press the dough to flatten and roll into rectangular shape of 6" x 12" by a rolling pin, arrange filling onto the dough evenly. Fold 1 cm from both sides inwards. Roll the dough from the top to the bottom. Make the tip thinner and press firmly.
5. Place the dough into a greased mold for final fermentation. (Refer to method of final fermentation on p.35)
6. Wait until the dough rises to 80% full, brush the surface with egg mixture. Pipe some salad dressing and sprinkle with chopped cheese.Bake in the lower deck of a preheated oven at 180°C for 15 minutes. Change to the middle deck. Cover with a sheet of aluminium foil and bake for 10 more minutes. Leave to cool and remove mold. Serve.

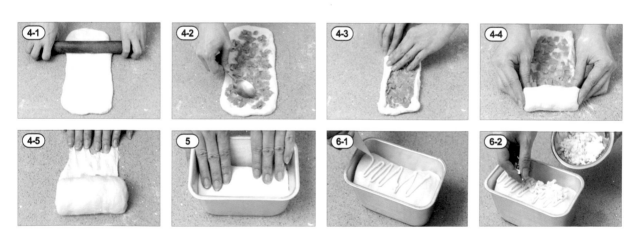

 小提醒 Tips

洋葱要預先炒至軟化，否則烤完會較生且有辛辣味。
Onion should be stir-fried till soft, or it will be raw and having spicy flavor after baking.

蒜香方形麵包
Garlic Bread

麵包模 Bread Mold

4吋×6.5吋長方形烤模
4" X 6.5" rectangular
baking mold

材料　Ingredients

高筋麵粉　250克/g
high gluten flour

酥油　20克/g
shortening

細鹽　5克/g
fine salt

奶粉　3克/g
milk powder

溫水　160毫升/ml
warm water

砂糖　15克/g
sugar

即溶酵母　4克/g
instant yeast

餡料　Filling

蒜蓉醬（製法請參閱p.22）
The method of arlic paste
（refer to p.22）

表面　Topping

白芝麻　少許
some white sesame seeds

全蛋液（刷表面）　適量
some egg mixture for glazing

做法　Method

1. 詳細的基本麵包搓揉方法及第一次發酵方法請參閱 p.27。

2. 麵糰平均分割成 2 份，拍摺成長形，讓麵糰鬆弛 15-20 分鐘。

3. 將麵糰按扁，用木棍桿薄成 6 吋 ×12 吋長形後，將麵糰由上向下捲起，收口略為拉薄並壓實。

4. 放入已刷油的烤模內，進行最後發酵。（最後發酵方法請參閱 p.35）

5. 待麵糰發酵至 8 成，在表面輕刷上全蛋液，灑上少許芝麻，放入已預熱的烤箱底層，用 180℃烤 15 分鐘，然後改放在中層，表面蓋上鋁箔紙再烤 10 分鐘。麵包出爐即脫模放涼。

6. 待麵包放涼再切片，每片麵包塗上適量蒜蓉醬，放回烤箱，用 150℃烤 5-8 分鐘即可。

1. For basic method of bread kneading and first fermentation, refer to p.27.

2. Divide the dough into 2 portions equally, roll into rods, let the dough rest for 15-20 minutes.

3. Press the dough to flatten and roll into rectangular shape of 6" x 12" by a rolling pin. Roll the dough from the top to the bottom. Make the tip thinner and press firmly.

4. Place the dough into a greased mold for final fermentation. (Refer to method of final fermentation on p.35)

5. Wait until the dough rises to 80% full, brush the surface with egg mixture. Sprinkle with some white sesame seeds.Bake in the lower deck of a preheated oven at 180°C for 15 minutes. Change to the middle deck. Cover with a sheet of aluminium foil and bake for 10 more minutes. Leave to cool and remove mold. Serve.

6. Slice the bread when it is cool, spread garlic paste onto each slice of bread. Bake in oven at 150°C for 5-8 minutes.

小提醒 Tips

如果麵包是留待隔天早餐時吃，蒜蓉醬可以不用先塗上，等到吃時一同烘烤，香味更濃。
Do not spread garlic paste if the bread is left until breakfast the other day, the flavor will even better if the paste is spread just before baking.

香芋番薯方形麵包
Taro and Sweet Potato Bread

麵包模 Bread Mold

4吋×6.5吋 長方形烤模
4 " X 6.5 " rectangular
baking mold

材料　Ingredients

高筋麵粉　250克/g
high gluten flour

酥油　15克/g
shortening

奶粉　8克/g
milk powder

細鹽　2克/g
salt

溫水　135克/g
warm water

砂糖　30克/g
sugar

雞蛋　20克/g
egg

即溶酵母　4克/g
instant yeast

蒸熟芋頭絲　100克/g
steamed taro shreds

糖粉（2個份量）　30克/g
icing sugar (2 servings)

表面　Topping

馬自瑞拉起士　適量
some mozzarella

番薯泥　100克/g
sweet potato puree

全蛋液（刷表面）　適量
some egg mixture for glazing

做法　Method

1. 詳細的基本麵包搓揉方法及第一次發酵方法請參閱 p.27 。
2. 將麵糰平均分割成 2 份，拍摺成長形，讓麵糰鬆弛 15-20 分鐘。
3. 將麵糰按扁，用木棍桿薄成 6 吋 × 12 吋長形後，將一份紫色麵糰放在一份白色麵糰上，加入芋頭絲及適量番薯餡，由上向下捲至底部，將收口壓實。
4. 將麵包放入已刷油的方形麵包模型內，進行最後發酵。（最後發酵方法請參閱 p.35）
5. 待麵糰發酵至兩倍大，表面刷上全蛋液，然後灑上起士碎粒，放入已預熱的烤箱底層，用 180℃烤 15 分鐘，然後改放在中層，表面蓋上鋁箔紙再烤 10 分鐘即可。

1. For basic method of bread kneading and first fermentation, refer to p.27.
2. Divide the dough into 2 portions equally, roll into rods, let the dough rest for 15-20 minutes.
3. Press the dough to flatten and roll into rectangular shape of 6" x 12" by a rolling pin. Place 1 portion of purple dough onto 1 portion of white dough, arrange taro shreds and sweet potato filling onto the dough evenly. Roll the dough from the top to the bottom. Make the tip thinner and press firmly.
4. Place the dough into a greased restangular mold for final fermentation. (Refer to method of final fermentation on p.35)
5. Wait until the dough rises 2 times in volume, brush the surface with egg mixture. Sprinkle with some chopped cheese. Bake in the lower deck of a preheated oven at 180°C for 15 minutes. Change to the middle deck. Cover with a sheet of aluminium foil and bake for 10 more minutes. Leave to cool and remove mold. Serve.

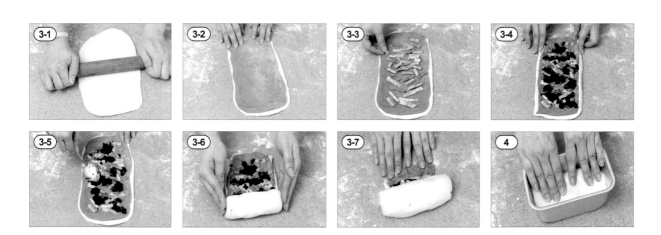

 小提醒 Tips

若不想加入紫心番薯，可加香芋香精進麵糰，搓揉成紫色。
Taro essence can be used instead of purple sweet potatoes to make the dough purple in color.

三色饅頭
Tri-color Steamed Bun

白色原味材料 6個份量
White Original Flavor
Ingredients (6 servings)

麵粉　160克/g
flour

泡打粉　1/4 茶匙/tsp
baking powder

酥油　10克/g
shortening

溫水　80毫升/ml
warm water

砂糖　30克/g
sugar

即溶酵母　1/2 茶匙/tsp
instant yeast

綠茶口味材料 6個份量
Green Tea Flavour
Ingredients (6 servings)

麵粉　155克/g
flour

泡打粉　1/2 茶匙/tsp
baking powder

綠茶粉　3克/g
green tea powder

酥油　10克/g
shortening

溫水　80毫升/ml
warm water

砂糖　30克/g
sugar

即溶酵母　1/2 茶匙/tsp
instant yeast

番薯口味材料 10個份量
Sweet Potato Flavor
Ingredients (10 servings)

麵粉　200克/g
flour

泡打粉　1/2 茶匙/tsp
baking powder

酥油　1茶匙/tsp
shortening

砂糖　30克/g
sugar

即溶酵母　1/2 茶匙/tsp
instant yeast

溫水　100毫升/ml
warm water

紫心番薯泥　80克/g
purple sweet potatoes

餡料　Filling

紫心番薯餡80-100克（約半
份的份量），半份有番薯
餡，半份沒有
80-100g purple sweet
potatoes filling (half
portion), half portion with
sweet potato filling and
half portion without sweet
potato filling

做法　Method

1. 將麵粉、泡打粉預先過篩，放入大鋼盆內。
2. 將砂糖、酵母、酥油及溫水加入材料內拌勻及搓揉成光滑麵糰。
3. 用保鮮膜蓋好，放室溫發酵約 20 分鐘。
4. 將麵糰取出，分均分成 2 份。每份用木棍桿成 10 吋 × 6 吋的長方形，平均塗上番薯餡料，然後慢慢由上向下捲成圓條狀，收口壓實。用塑膠刮板將麵條分割成 5 份（頭尾切掉），底層墊上烘焙紙，放入蒸籠內進行最後發酵 20-30 分鐘。
5. 將已發酵的饅頭放入蒸鍋內，大火蒸約 8 分鐘即可。
6. 重複步驟以其他材料製作不同味道的饅頭。

1. Sieve flour, baking powder into a large basin.
2. Add sugar, instant yeast, shortening and warm water into the flour, mix well and knead into a smooth dough.
3. Cover with plastic wrap, ferment for 20 minutes at room temperature.
4. Take out the dough and divide into 2 portions. Roll each portion into a 10" x 6" rectangle by a rolling pin. Arrange sweet potato filling onto the dough evenly. Roll into a circular strip downwards slowly, press the opening. Divide the dough into 5 equal portions (cut off both sides). Arrange onto a steamer lined with butter paper for the final fermentation for 20-30 minutes.
5. Steam the buns in wok over high heat for about 8 minutes and serve.
6. Repeat by using different ingredients to make steamed buns in different flavours.

  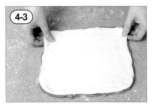 
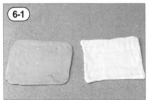  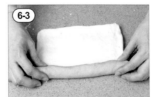 
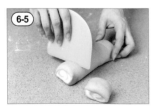 

 小提醒 Tips

1. 關火後要立即將鍋蓋拿起，否則凝結的水氣會滴在饅頭表面，影響外觀。
2. 麵糰頭尾切掉一小節，做出來的外觀效果更佳。

1. Remove the lid of the wok immediately after the heat is turned off or the running liquid from the lid of the wok will flow back and affect the appearance of the bun.
2. Cut off both sides of the dough will improve the appearance.

小麥胚芽芝麻花捲
Wheat Germ and Sesame Floral Roll

材料　Ingredients

麵粉　140克/g
flour

酥油　1茶匙/tsp
shortening

芝麻粉　25克/g
sesame flour

小麥胚芽　5克/g
wheat germ

溫水　85毫升/ml
warm water

砂糖　30克/g
sugar

即溶酵母　2克/g
instant yeast

表面　Topping

黑芝麻（已炒香）　適量
some black sesame
seeds (stir-fried until
fragrant)

做法　Method

1. 將所有材料拌勻並搓成光滑麵糰，蓋上保鮮膜，放室溫發酵 20 分鐘。
2. 將麵糰用木棍桿成約 13 吋 × 6 吋 的長方形薄片。
3. 將兩邊薄片向中央摺疊成 3 層，然後將麵皮拉約 8 吋 × 8 吋方形。
4. 將麵皮平均切成細條，拿起 3-4 條捲成花捲，在表面噴少許水，灑上少許黑芝麻，將剩下的麵皮全部依此方法完成。
5. 將花捲墊上底紙，放入蒸籠內，放在和暖的地方發酵 20 分鐘。
6. 發酵完成後，用鍋大火蒸約 8 分鐘即可。

1. Mix all ingredients and knead into a smooth dough, cover with plastic wrap, rest in room temperature for 20 minutes.
2. Roll the dough into a thin rectangle of about 13" X 6" by a rolling pin.
3. Fold both sides of the thin dough to the centre into three layers, then stretch the dough into about 8" x 8" square.
4. Cut the dough into thin strips evenly, pick up 3-4 strips and roll into floral shape. Spray some water, sprinkle some black sesame seeds. Finish the remaining dough according to this method.
5. Arrange onto a steamer lined with butter paper for the final fermentation for 20 minutes.
6. Steam the buns in wok over high heat for about 8 minutes and serve.

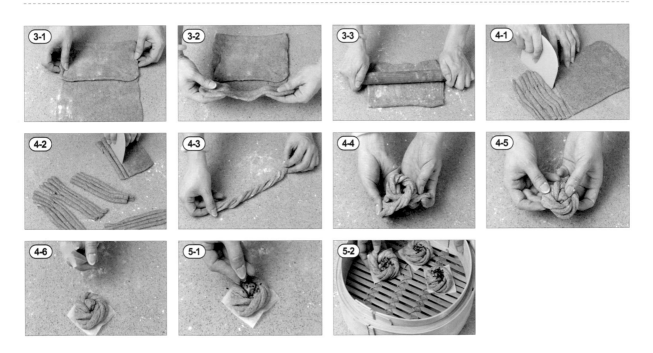

小麥胚芽含豐富維他命及蛋白質，纖維成分高，有助小腸蠕動。
Wheat germ is rich in vitamins and protein with rich fiber, and to help the peristalsis of intestines.

奶黃包
Egg Custard Bun

材料　Ingredients

麵粉　150克/g
flour

泡打粉　1茶匙/tsp
baking powder

即溶酵母　2克/g
instant yeast

砂糖　30克/g
sugar

酥油　8克/g
shortening

溫水　85毫升/ml
warm water

餡料　Filling

奶黃餡　120克/g
（請參照 p.26 的製法）
egg custard filling
(refer to p.26)

做法　Method

1. 將麵粉及泡打粉預先過篩並放入碗內，然後將即溶酵母、砂糖、酥油及溫水加入搓匀成麵糰（約揉 5-8 分鐘）。放回入碗內蓋上保鮮膜，發酵 20 分鐘。（第一次發酵）
2. 將麵糰平均分成 6 份，包入一份奶黃餡，黏緊收口，將麵糰反轉收口向下，墊上底紙或烘焙紙，放入蒸籠進行最後發酵 20 分鐘。（蒸籠放在廚房或溫暖的地方）
3. 預先在鍋內將水煮至大滾，將蒸籠放在鍋上，用大火蒸約 8 分鐘即可。

1. Sieve flour and baking powder into a bowl, then add instant yeast, sugar, shortening and warm water and knead into a dough evenly. (Knead for about 5-8 minutes). Put the dough back to the bowl, cover with plastic wrap and ferment for 20 minutes. (First fermentation)
2. Divide the dough evenly into 6 portions, add 1 portion of egg custard filling, press the opening firmly. Invert the dough and place the openings at the bottom. Arrange onto a steamer lined with butter paper for the final fermentation for 20 minutes. (Place the steamer in the kitchen or a warm place.)
3. Bring a wok of water to a boil. Steam the buns in wok over high heat for about 8 minutes and serve.

小提醒 Tips

奶黃餡蒸完放涼後，一定要將餡料用手搓揉或壓平至光滑，吃時才柔軟。
After steaming egg custard filling and leaving to cool, it should be rubbed or pressed and flattened until smooth and soft.

第一次學做手工麵包！超安心

作　　　者　韋太
發　行　人　程安琪
總　策　劃　程顯灝
執行編輯　譽緻國際美學企業社、盧美娜
主　　　編　譽緻國際美學企業社、盧筱筑
美　　　編　譽緻國際美學企業社
封面設計　洪瑞伯

出　版　者　橘子文化事業有限公司
總　代　理　三友圖書有限公司
地　　　址　106 台北市安和路2段213號4樓
電　　　話　（02）2377-4155
傳　　　真　（02）2377-4355
E-mail　service@sanyau.com.tw
郵政劃撥　5844889　三友圖書有限公司

總　經　銷　大和書報圖書股份有限公司
地　　　址　新北市新莊區五工五路2號
電　　　話　（02）8990-2588
傳　　　真　（02）2299-7900

http://www.ju-zi.com.tw

初　　　版　2014年02月
定　　　價　新臺幣 298元
I S B N　978-986-6062-69-8（平裝）

國家圖書館出版品預行編目(CIP)

第一次學做手工麵包!超安心 / 韋太作. -- 初版. --
臺北市：橘子文化, 2014.02
　　面；　公分

ISBN 978-986-6062-69-8(平裝)

1.點心食譜　2.麵包

427.16　　　　　　　　　　102026185

SANYAU
三友圖書 / 讀者俱樂部

填妥本問卷，並寄回，即可成為三友圖書會員。
我們將優先提供相關優惠活動訊息給您。

優質好康

粉絲招募
歡迎加入

○ 看書 所有出版品應有盡有
○ 分享 與作者最直接的交談
○ 資訊 好書特惠馬上就知道

旗林文化╳橘子文化╳四塊玉文創
https://www.facebook.com/comehomelife

親愛的讀者：
感謝您購買《第一次學做手工麵包！超安心》一書，為感謝您的支持與愛護，只要填妥本回函，並寄回本社，即可成為三友圖書會員，將定時提供新書資訊及各種優惠給您。

1 您從何處購得本書？
□博客來網路書店 □金石堂網路書店 □誠品網路書店 □其他網路書店
□實體書店_____

2 您從何處得知本書？
□廣播媒體 □臉書 □朋友推薦 □博客來網路書店 □金石堂網路書店
□誠品網路書店 □其他網路書店_____ □實體書店_____

3 您購買本書的因素有哪些？(可複選)
□作者 □內容 □圖片 □版面編排 □其他_____

4 您覺得本書的封面設計如何？
□非常滿意 □滿意 □普通 □很差 □其他_____

5 非常感謝您購買此書，您還對哪些主題有興趣？(可複選)
□中西食譜 □點心烘焙 □飲品類 □瘦身美容 □手作DIY
□養生保健 □兩性關係 □心靈療癒 □小說 □其他_____

6 您最常選擇購書的通路是以下哪一個？
□誠品實體書店 □金石堂實體書店 □博客來網路書店 □誠品網路書店
□金石堂網路書店 □PC HOME網路書店 □Costco
□其他網路書店_____ □其他實體書店_____

7 若本書出版形式為電子書，您的購買意願？
□會購買 □不一定會購買 □視價格考慮是否購買 □不會購買
□其他_____

8 您是否有閱讀電子書的習慣？
□有，已習慣看電子書 □偶爾會看 □沒有，不習慣看電子書
□其他_____

9 您認為本書尚需改進之處？以及對我們的意見？
_____

10 日後若有優惠訊息，您希望我們以何種方式通知您？
□電話 □E-mail □簡訊 □書面宣傳寄送至貴府 □其他_____

謝謝您的填寫，
您寶貴的建議是我們進步的動力！

姓名_____　　出生年月日_____

電話_____　　E-mail_____

通訊地址_____